Curious Countries

Lars Thomas

Typeset by Jonathan Downes,
Cover and Layout by SPiderKaT for CFZ Communications
Using Microsoft Word 360; Microsoft Publisher 360; Adobe Photoshop CS.
Cover illustration by Jacob Rask
Eyewitness reconstruction pictures by Linda Kjaer-Thomsen

First published in Great Britain by CFZ Press

CFZ Press
Myrtle Cottage
Woolsery
Bideford
North Devon
EX39 5QR

© CFZ MMXVIII

ISBN: 978-1-909488-57-1

To Jeanett, who should have been here...

Contents

INTRODUCTION – *Folklore, cryptozoology and zoology*

The countries of Northern Europe are a strange and diverse lot. Bleak and inhospitable in some places, mild and gentle in other areas, intensely built up or nothing but bloody pine forests. People have been around for thousands of years, giving most of us northerners a profound sense of history, and a rather detailed knowledge of our homelands.

From long before the Viking ages, we have cast a wary eye towards the deep dark forests, the bleak and inhospitable mountains. These were places to be careful about. You could get lost, fall off a cliff, or even meet the locals. Wolves, moose and bears roamed these parts, but so did other creatures – big, small, friendly, dangerous, and sometimes utterly terrifying. This after all is the land of Fenrir, the giant wolf of old Norse mythology, and the former haunts and killing ground of King Beowulf (a Swede), who travelled to Denmark to kill the terrible monster Grendel who was giving King Hrothgar a seriously hard time. These creatures may only exist in folklore and the imagination of any scared and bewildered child, but then again – they may also be real.

This is the companion volume to my earlier book *Weird Waters* (CFZ Press 2011) about aquatic monsters in Northern Europe and the Baltic countries. *Curious Countries* is a book about all the strange creatures that roam the dry land, bury beneath the ground, and fly through the air. Some are benevolent, some most definitely are not.

Northern Europe and the Baltic states are not, alas, located in the tropics, something which is abundantly clear as I write this on a cold, grey, wet and horrible day in the middle of winter. So, our natural wildlife and plant life is not particularly rich and varied, compared to, say, the Amazon rainforest, although it is quite nice and interesting.

But should we turn to the unnatural life, we are in a part of the world which in no way lacks anything in the department of the strange and mysterious. We do have a very rich and varied folklore and sightings aplenty of all kinds of strange creatures and mysterious beings.

There are heaps (or is it hordes?) of out-of-place animals – well known species as it were, but cropping up in places where they have absolutely no business to be – big cats (of course), kangaroos, crocodiles, even a cassowary that washed up dead on a Danish beach a number of years ago.

We have a few out-of-time animals; sightings of wolves in Denmark several centuries after the killing of the last official one, to name but one. However, things in that department have changed suddenly over the last few years, with the appearance of several wolves in Denmark, and the fact that they have now actually bred here for the first time in more than 200 years. And then of course there are the truly strange ones, farting dragons (pardon my French), giant horse-eating lizards and a truly staggering variety of human-like beings in all shapes and sizes, from various forms of the little people to some rather grim and gruesome examples of rather large people; trolls, giants and such likes.

There are even stories about some of the old Scandinavian kings being expert troll-hunters – or rather hunters of large, hairy, bipedal humanlike creatures with long floppy breasts (sounds familiar?).

We also have strange insects, flies with glowing heads and talking beetles, a bird of prey almost as big as the legendary Roc, and a few even more bizarre odds and ends.

I have always to some extent felt, that Bernard Heuvelmans actually did the whole study of cryptozoology a disservice by making it a separate field of study. To me, cryptozoology has always been - and will always be - a natural continuation of zoology. In a sense, 200 or 300 years ago, *all* zoology was cryptozoology. And likewise - at the other end of the spectrum - the study of folklore and folkloristic beasts is just as much an extension of zoology; if nothing else zoology must have supplied the inspiration for many of the more fantastic beasts. So – as I see folklore, cryptozoology and zoology as the subjects, that merge naturally, this book will contain creatures from all parts of the spectrum, and they will be treated (I hope) equally.

I should perhaps point out – before any reader ventures any further into this tome - that I am not personally a big fan of reading sighting after sighting described in brain-numbing detail in a book – reading stuff like that is what you do *before* you write the book. The only sightings you will see described in any detail in here are the ones that truly *add* something to our understanding of whatever creature we are discussing. So, the style of this book is perhaps a bit different from quite a number of other cryptozoological books, but hopefully still entertaining.

As usual I should be thanking an awful lot of people, but instead of making a long list, and probably forgetting some anyway, I'll just extend my thanks and gratitude to all those people who have told me their stories during the last three decades, or shared their own findings with me. Without you and your patience, this book could never have been written.

But enough of these boring formalities – let's get on with it. Enjoy - and keep your eyes open, and your camera ready!

And should you hear an interesting story – please let me know. I can be contacted through the publishers, through my facebook page or my blog The Cryptodane, where you will also find snippets of cryptozoological news – mostly from Denmark, but also from other parts of Scandinavia if something catches my imagination.

Copenhagen 2018
Lars Thomas

GREENLAND – Monsters of the very far north

f you have an inflated sense of your own ego, I can heartily recommend a trip to Greenland. Never in my ramblings all over this strange little planet of ours, have I seen a landscape so magnificent, and never have I felt so small and insignificant. It is most refreshing, and most efficient for cutting people down to size. To say that the Greenland landscape is beautiful and majestic is to indulge in an understatement of epic proportions. It is quite simply out of this world. Standing on top of a glacier in West Greenland you feel like you can see to the end of the world, and thanks to the cold dry air, you almost can. It is also a place, where you should keep your wits about you at all times. Up here the nature can and will kill you, and it will play terrible tricks on your senses. In the middle of summer, the sun never sets, and that kind of plays havoc with your internal rhythm, and that cold clear air again can make it extremely difficult to judge sizes and shapes. Things that are very close can seem far away, and vice versa.

Greenland is a bit desolate animal wise. In fact that is not entirely true – the sea surrounding Greenland absolutely teems with life. On dry land though things are a bit more sparse. The actual number of species is quite small. Only a handful of insects, no reptiles or amphibians, a fair number of birds though, but only a few mammals. One of the most visible ones is the arctic fox; a small, curious, bold and tough little rascal that manages to eke out an existence in even the bleakest of landscapes, and indeed in the middle of winter, being the lucky owners of one of the densest and warmest furs known.

I can from personal experience also attest to the fact, that keeping food and butter in your tent in no way keeps the arctic foxes from stealing it – or your candles, socks, t-shirts and any other even remotely edible looking thing. And the fact that the actual number of insect species is small is no guarantee that the number of individuals won't be vast. Living in a place where even the animals are struggling to survive, the indigenous people of Greenland have developed unparalleled skills in hunting and fishing, and an intimate knowledge of their surroundings and the plants and animals with which they share the narrow strip of land around the margins of the enormous Greenland ice cap. If anything strange lives here – they would know.

Greenland is in fact not particularly well endowed when it comes to monsters and other mysterious creatures of the terrestrial persuasion. This may have something to do with the fact that Greenland is particularly lacking in deep dark forests for these beings to disport themselves in. I have been in several Greenland forests, but as the tallest and most common species of tree is arctic willow, which only manages to grow to something like 60 cm in height, it is not something you lose yourself in easily – unless you crawl on your stomach. Further north it is even more difficult. When the Arctic dwarf birch takes over, you would have to be the thickness of an envelope to hide anywhere.

There is of course the enormous freezing cold and utterly bleak ice cap, but I don't think anything, monster or otherwise, could live there for any lengthy period of time. But basically, if something out of the ordinary is living anywhere in Greenland, the local people would have seen it a long time ago – and in most cases, they haven't, so...

But that doesn't mean Greenland is completely devoid of interest, far from it. Ancient legends as well as a small number of sightings, does give some indication of the existence of beings quite a bit out of the ordinary.

Feathered fiends
The most visible and dominating group of animals in Greenland are the birds (not counting the most obnoxious ones, which are by far the mosquitoes, but that's a whole different matter – 154 mosquito bites on an area of shoulder

roughly the size of a palm indicates one forgot to close one's tent properly).

Greenland birds are a rather diverse lot, from small colourful Lapland buntings to awe-inspiring creatures like the Greenland white-tailed sea-eagle, an enormous beast with the wingspan of a couple of barn doors, and a face armed with an enormous beak looking like a giant can opener capable of wreaking bloody havoc should the need arise. And don't talk about the claws – they make me shudder every time I see them up close.

The Greenland birds are by and large a rather conservative lot – usually you don't see much change in the composition of the local birdlife, but every now and then strange things do happen. Despite their highly developed senses and their magnificent powers of navigation, birds do get lost in a fairly regular manner. Something every birdwatcher knows, and every twitcher gets high on every spring and autumn. Occasionally a rare visitor from North America or some such place turns up in Greenland as well, but usually to be met by a fairly modest reception committee, as birdwatchers and twitchers are few and far between up here.

But every now and then some truly weird ones show themselves – birdlike creatures completely unlike anything you normally find among the local birdlife. Such as small, and extremely colourful birds with magical abilities, and beings so large, that you automatically start to think along the lines of North American thunderbirds and similar monsters. These being are extremely rare, but strangely enough they are often associated with storms from unusual directions. Thunderstorms are rare in Greenland, but there are plenty of ordinary storms, and then you do not want to be out and about, let alone sailing. Take it from me, I've been out there, and it is not funny!

The giant gull
Buried deep within the bowels of Greenland folklore is a strange story about an enormous white gull that turns up every now and then, and generally scares the pants of everybody. It is large enough to carry off a full grown person, usually depositing said person on the top of a mountain located somewhere suitably bleak and remote. And it is an unusually brave and resourceful person that manages to find his or her way back to their loved ones.

This of course sounds very much like a fairy-tale, and perhaps it is. Just another birdlike monster designed to scare children and keep them away from dangerous places. But then again – there might just be some form of reality behind the myth. You see I have personally talked to two people who claim to have seen the giant gull, one of whom was, if not carried away by the gull then at least attacked by it.

Greenland is of course already home to a couple of white gulls, the glacous and the Iceland gull. As gulls go, they are fairly big ones, but essentially within normal gull parameters. The biggest, the glacous is the size of a common herring gull. But the giant gull is truly enormous, variously described as having a wingspan from between two and four or perhaps even five metres and sometimes even more. In one memorable story a man described it as having a body the size of a beluga whale, and a corresponding wingspan, which would be somewhere in the vicinity of 10 metres.

The first sighting took place in 1926. A fisherman in a little boat of the southern tip of Greenland was tending to his nets, when some kind of movement he just caught out of the corner of his eye, made him look up. At first he couldn't understand what he was seeing. It was definitely a bird of some kind, but bigger than anything he had ever seen. It was flying along in big lazy circles a couple of hundred metres away. The bird was mainly white, but the back and over side of the wings was very dark, almost black, and it had a very large beak. But the scary thing was the size. The bird never came any closer, so it was difficult for the fisherman to ascertain how big it was, but the fisherman thought it had a wingspan as least as long as his boat, which would put it somewhere around the seven metre mark. A truly enormous bird by all accounts.

The fisherman had of course heard the stories of the giant gull, and was deeply frightened, but the big bird paid him no heed at all and just kept flying past him until it disappeared in the distance. Deeply shaken the fisherman set out for the safety of dry land. His granddaughter, who told me the story some 60 years later, could still remember how pale and shaken he was, when he came home.

As the second sighting shows not even dry land is safe from the great gull. Sometime at the end of the summer of 1935 following a period of strong south easterly winds, a boy named Jakob, who at that time was 14 years old, and who lived in Manitsoq on the west coast of Greenland, was walking along the coast at low tide. The tidal difference at Manitsoq can be quite large, 6 metres is not uncommon, and Jakob was hoping to find something interesting and useful exposed by the retreating waters. He was making his way along the coast, when he suddenly saw an enormous shadow pass over him, and when he looked up, he saw an enormous white bird passing over his head, and then turning in a big lazy circle over the sea before returning in what Jakob perceived as a determined attack. He had never seen anything like this enormous white bird, so he started running home as fast as he could, and in what might best be described as blind terror. He later swore that the bird passed over him several times, low enough for him to feel a rush of air through his hair. When he reached his parents house the bird had gone, but when he told me the story – I met him in Søndre Strømfjord in 1984 – the memory still made him jumpy.

The third sighting that I know of is nowhere as dramatic as the second one. It took place sometime in the early 1960's. This story only comes to me second hand. It was told to me by a half Greenlandic half Danish zoology student,

who heard it from the father of one of her fellow and full-blooded Greenlandic students. When he was a young boy living in Sisimiut on the West Coast of Greenland he had seen an extremely large bird sitting on a rock just outside the harbour in Sisimiut. It looked like an enormous gull, but with a very big black beak with a yellow stripe down the middle. Nothing dramatic happened. He just went home and told his father about the big bird, fully expecting the father to get his gun and shoot the bird, but his father simply refused to believe him, and nothing more was said about it.

This all sound rather strange, but it was in fact the last of these stories that made me wonder. Could we in fact be talking about rare sightings of albatrosses? I mean, they do look somewhat like giant gulls, and they are so big that suspecting them of being able to carry children or perhaps even adults away are not that crazy. In august 1935 a black-browed albatross was in fact shot at Lille Hellefiskebanke, which is located of the coast at Manitsoq – perhaps the bird that "attacked" Jakob? And in 1963 a yellow-nosed albatross was caught in Greenland waters as well. And that particular species does in fact have a black bill with a yellow stripe down the middle. So, a bit of southerly or south easterly winds, and hey presto – there be albatrosses around Greenland. I know that albatrosses normally are only found in the southern Atlantic, but a few foolhardy individuals venture north every now and then. One particular memorable individual spent many years in a gannet colony in the Shetland Isles. Maybe that bird took a swing around Greenland every now and then...

The weatherbird

Mysterious and unknown giants are of course always a lot more exiting and dramatic than smaller ones, but one should not scorn the small – it can be so much more interesting. That could be said for a trio of strange birds that apparently show their rather coloured self in Greenland every now and then. These are all small birds, as I've said, nothing like the giant we have just discussed, but they are apparently peaceful and benevolent – not dangerous or anything, but apparently gifted with some rather awesome powers despite their diminutive size. Oh yes, and some rather startling colours. I only have a handful of sightings of these birds, but despite them looking nothing like the giant gull – or albatross – I suspect we are dealing with a similar phenomenon.

The smallest one of the lot is only "the size of a gull's egg", but don't let that fool you. It has magical properties – it can disappear from one place, and in an instant appear again in another place. It doesn't need wings to fly with, and what is perhaps the most important thing – it brings warm and sunny weather from the south.

Now this all sound like something from one of Hans Christian Andersen's fairy tales, but I think there is a fairly straightforward explanation. If we start

from the back – warm and sunny weather from the south (or perhaps southwest): That would be from North America. Do we in fact have any birds in North America that are very small, would appear to disappear in one place and appear in the next within a blink of an eye, and birds that can also, apparently, fly without using its wings? Oh no, we haven't I hear someone cry in the distance, to which I can only reply: Oh yes, we do! A hummingbird and more precisely, the rubythroat which can be found all the way up to south-eastern Canada. Enter some warm weather, and a good wind blowing north towards Greenland, and I see no reason why the odd hummer or two shouldn't be blown all the way to the shores of Greenland. They probably wouldn't last long, but every now and then they might live long enough to be seen by a surprised hunter or fisherman going about their daily business.

The bloodbird and the skybird

The same thing can probably be said about the final two feathered funnies, the bloodbird and the skybird. Nothing special about them, except for the fact, that the bloodbird is as red as a splash of blood from a newly shot seal, and the skybird looks like a tiny piece of the sky has sprouted wings and started to fly.

Both of these creatures are apparently extremely rare, I have only managed to collect two sightings of each, but they are again associated with winds from the south.

The people who saw them knew nothing about any special abilities the birds might have or anything like that. They only noticed their very strange colours. And they would indeed stand out in a Greenland landscape, but I suspect it is a case, or several cases as it were, of red cardinals and bluebirds from North America being blown way out of course and ending up in the Greenland wilderness.

Of mice and men... and bears

The number of land-living mammals in Greenland is limited as one would expect. There are a few big species, reindeer, muskoxen, and of course polar bears, although I suppose whether the polar bear is actually a land mammal is debatable. There are wolves here and there, and the ubiquitous arctic fox, as well as a few smaller species, chiefly lemmings. There is nothing special about any of these species. They are all typical arctic animals, perfectly adapted to a life in these harsh surroundings. But – and there is always a but – there have also been sightings of an animal that looks like a giant short-tailed mouse or perhaps an enormous guinea pig.

Had we been somewhere in South America I wouldn't have thought twice about sightings like that – capybaras and pacas could easily mascarade as giant mice or giant guinea pigs, but in Greenland? I have no idea what we are dealing with here, but I have collected altogether seven sightings dating from 1936 to 1981 all from the southern part of the west coast of Greenland from Søndre Strømfjord to Kap Farvel. All the sightings describe an animal looking like a very large lemming, with short legs, very short or non-existing tail, and very short rounded ears, all typical traits of an arctic animal. The only difference is that these "super" lemmings are about one metre in length, some 8-10 times longer than a normal lemming.

What? Are we really dealing with an enormous rodent, or is there some kind of logical explanation for this? As I mentioned above, I have no idea. I can just about imagine a Dane confusing perhaps some kind of seal on land with a big lemming, but that's about it. I am very much open to suggestions.

The big bear
The polar bear is as I have mentioned above, only a part-time land mammal, and it is by far the largest land living creature in Greenland – officially that is, because I do have a few sightings of another, and according to some stories even bigger bear, although it looks nothing like the brown bear which I suppose is the only species that could conceivably show up in Greenland. All the sightings I have gathered about this fearsome beast (17 all together) come from two areas in Greenland. One group of sightings (6) are concentrated around Tassilaq in south-eastern Greenland. They came to me many years ago courtesy of an engineer who worked in a company where I worked whilst studying at the University of Copenhagen. I took care of the company's considerable population of photo-copiers, and in a desperate effort not to bore myself to death I started talking to this engineer, whose name was Holger. Just like me he was interested in folklore, and whilst living and working in Tassilaq for some years he had gathered various interesting stories and tall tales, and we spent many hours – far too many according to the boss – discussing Greenland and various other exotic places we had visited.

Among his many stories were some told to him by various trappers and hunters in the area about an enormous and very strange bear that sometimes turned up very late in winter or early spring, if the winter had been especially cold. Now Greenland winters are no laughing matter at the best of times, especially in eastern Greenland, so a very hard and cold one hardly bears thinking about. But such were the words of the old timers he got to know.

I'll get to the actual description of the bear in a moment, but first we have to make a detour to extreme north-eastern Greenland, from where I have 11 more sightings of the same kind of bear. These sightings was given to me by a former member of the so-called Sirius Patrol, an elite military patrol whose main task is to maintain Danish sovereignty in this rather far-flung corner of the world. The members of the patrol are out and about even in the middle of winter, with their sledges and sledge-dogs and sometimes they see things. Niels, as his name is, looks very much like a bear himself, albeit a rather gentle one, but nevertheless he claimed that he and several of his friends and colleagues at various times had seen a bear that was definitely NOT a polar bear. All the sightings from his part of Greenland are from the 60's and 70's whereas the sightings from the area around Tassilaq are all from the 1930's and 1940's.

Get to the point already, I can hear people shouting in the back row – and how right you are. What have these people actually seen? Well, it's a bear, but a very strange one. First of all it is not interested in people in any way. Polar bears can be extremely curious, but this one takes absolutely no notice of people. Admitted all the sightings are from a distance, sometime a fairly large distance – in one case almost a kilometre, although the animal was seen through binoculars.

The bear is at least the size of a polar bear and sometimes even larger. The fur is white or with a yellow tint just like polar bears, but it can also be grey or a very light brownish colour. The main difference though is the animal's very long legs. Polar bears have rather short and stumpy legs, and thus a rather low-slung body, but this thing has very long legs giving an impression of being able to run very fast if necessary. The head is bigger and more squarish than the polar bears rather small and pointed head, and the ears are bigger. None of the animals have been seen hunting or doing anything else than just passing

Compare the polar bear (below) with the artist's impression of the Greenland 'Big Bear'

Domesticated reindeer

by, so I have no information as to biology or behaviour, but this rather brings to mind the *irkutiem*, the long-legged bear of Siberia. I suppose it is conceivable that bears from Siberia could cross the Bering Strait, but that still doesn't answer the question as to what this bear actually is – some special colour variation/mutation of a polar bear, a hybrid between polar bears and other species of bears, or a completely different species? All I can say is that if a Greenland hunter ever shoots one of these creatures, I hope he saves a sample or two for me.

Oh deer (sorry!)

Greenland is of course home to a rather well known species of deer, especially if you are of a certain childish disposition, or simply love Christmas. I am - of course - talking about the reindeer. You can see them in various parts of west Greenland, especially in the south, although there is population in the extreme northwest in the Thule area. They are well known animals, and no one would probably confuse them with any other species of deer. That is why a couple of sightings of a rather strange looking deer from the northwest of Greenland are so very interesting.

As deer go, reindeer are not terribly big, although the Greenland ones are a bit bigger than animals living further south in more gentle climates. Big males can end up somewhere in the neighbourhood of 200 kg. and in rare cases even more. They are rather short-legged and low to the ground, which make them seem even smaller. On the other hand they have quite enormous antlers, which to me always makes them look like they have tried to wear their father's headgear.

So what are we to make of this animal, seen by my correspondent Peter in 1984, when he was working at Thule Airbase:

> *"It was sometime in the autumn. For some reason I never did write down the actual date, but probably some day in late August or perhaps early September. I had been out trying to do a little bit of hunting not far from Qanaaq with my friend Andrew, one of the Americans at the base. We hadn't really seen anything, apart from a reindeer almost a mile away, and were on our way back when Andrew suddenly started yelling like a madman. We had passed a*

small depression in the landscape, and I had lagged a bit behind him. He was standing a bit higher up than me, and he was extremely agitated. When I got up to him, he was pointing ahead to an animal running away from us. It was a couple of hundred metres away, but my God it was big. It was the biggest deer I had ever seen. It was at least twice as big as the biggest reindeer either of us had ever seen, and I am quite certain it was even bigger than a moose. I have seen moose in Sweden, and I am quite certain it was not a moose. It looked like a gigantic red deer. But the most terrifying thing was the antlers. They were absolutely enormous. They looked like to enormous hands, but I think they had a span of at least 2 metres. Although both Andrew and I had a rifle, we completely forgot to use it. For some reason we both had a strange feeling that trying to shoot this enormous animal would be a waste of time. The deer kept running away from us at a steady pace, and in a fairly short time it had disappeared from view. We thought about following it, but decided not to, and just kind of stopped talking about it."

A similar animal, perhaps the same one, was seen three years later under almost identical circumstances. Peter had left the airbase by then, but his friend Andrew was still there and he saw the giant deer a second time.

What in God's or any other name can this be? I suppose a moose could go walkabout from somewhere in Canada and end up in Northern Greenland, but Peter was certain the animal was considerably bigger than a moose, which basically would mean an animal very close to a ton in weight or perhaps even more. It could of course have been an extremely large moose, but on the other hand Peter's description does not really support this. The animal reminded him of an enormous red deer, and a moose does not, even by the farthest stretch of the imagination, look like a red deer, giant or otherwise. There is one animal that does, as a matter of fact. I am not saying Peter saw one of these, but it does give one food for thought.

Behold the Irish elk (*Megaloceros giganteus*). This rather hefty animal used to roam over large parts of Eurasia until some 7-8000 years ago. It had a shoulder height of at least 2 metres, and a set of antlers measuring up to 3.7 metres from tip to tip, far larger than any other species known. The name comes from the

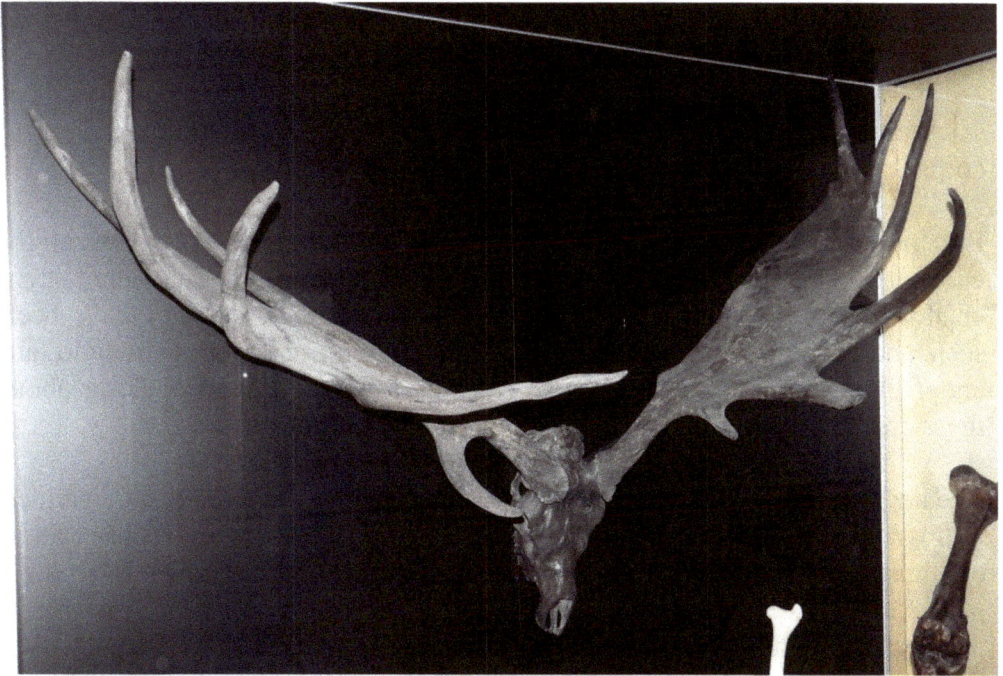

fact that a large number of specimens have been found in Irish peat bogs. Now again, I am not suggesting, that there is, or was, a relict population of Irish elks in northern Greenland or perhaps northern Canada, but the resemblance is quite uncanny. And just to add insult to injury, I also know of a couple of sightings of this magnificent beast from the other end of "my" area, in Eastern Estonia, around the turn of the 18/19th century. These animals were also seen trotting away at a fair distance, but even here the eyewitnesses were certain they were far bigger than even a moose.

The man-eaters
One might quite reasonably suggest that Greenland was an obvious place for a snowman, abominable or otherwise, but in fact things like BHM's are something of a rarity in Greenland. But they are out there, treading the thin line between fantasy and reality. One might even dare to call it "The Twilight Zone" (cue sinister theme music here!). Anyway, Greenland folklore has a number of legends and stories about humanlike creatures or perhaps ghoulish humans with a more than passing resemblance with yetis, bigfoots and similar beings from other parts of the world.

They are definitely humanlike, usually big and hairy, and walking on only two legs.

They are very smelly – in some cases smelling like a dead and decomposing corpse.

But they also have a rather sinister tendency to eat any humans they might encounter. This would call for a rather strained relationship, and I dare say a chance meeting would tend to ruin your day. Nevertheless – they have been seen, and are still being seen to this day, although it is rare for anybody to talk about them, possibly because some legends says that anyone talking about these creatures will be the next to fall victim to their hunger.

The approximately 50 sightings I have collected over the years can be divided into several distinct groups depending on weather and observation circumstances. In the following I will give an example of each one:

The most common form of sighting is a big lumbering shape usually seen during a snowstorm. In some cases it looks almost like a big moving boulder, in other cases it is a much more humanlike shape, standing some 3-4 metre in height, usually standing still, turning a smallish head from side to side before disappearing into the whirling snow again. If the weather is not too bad, you can sometimes hear a snuffling or growling sound, but usually the creature is silent. Sometime you catch a whiff of a serious case of halitosis mixed with a bit of death and decay.

This all sounds very dramatic, but we are after all in the land of the polar bear, so I am quite convinced that this is sightings of said species during difficult conditions. The big bears are quite capable of standing on their hind legs for some time, presumably in an effort to get a better view of their surroundings, and look around, perhaps sniffing their surroundings and growling a bit. And like all carnivores their breath is not exactly minty fresh, especially not if they have been feasting on the rotting carcass of a whale.

A smaller and less smelly type of upright creature is also typically seen during snowstorms. These creatures are always walking on their back legs, are slightly smaller than the first type, but don't seem to be interested in their

LKT
2 o18

surroundings in any way. In some cases they have even led people who were lost in the snowstorm back to safety, and then usually disappearing without a trace.

It could of course be other humans, lost in the snowstorm as well, but there is also a strange ghostlike quality to them, so perhaps they are something else entirely, tupilaks perhaps, a strange kind of monsters made by magicians and sent out to kill specific people, so perhaps being more benevolent towards the people who were not their specific targets.

In a few cases these beasties have been seen in fog, apparently moving along inside the fog banks following and mimicking every move of the people who have seen them. In these cases I am certain they have been some kind of reflection of distorted shadows of the people passing by. Several of the eyewitnesses to these phenomena have told how the creature in the fog would raise one arm if they did, and so on.

And then there are the very strange and very interesting ones. I only have three of those, but I think they are worth a closer look.

June 6th, 1976, about 10 km north of Grønnedal Naval Base in South-western Greenland. A fine day, hardly any wind, sunshine, and a temperature of some 15 degrees. Two men on leave from the naval base had gone out for a spot of fishing. They were walking merrily along, when their noses were suddenly assaulted by a stench so eye-wateringly bad, that both of them started retching and coughing. It was in a middle of a coughing fit, that one of them realized they were not alone. On a small hillock about 15 metres away an enormous hairy "gorilla-like thing" was standing and looking straight at them.

> *"I am certain it was intelligent – it was definitely sizing us up and trying to decide what to do with us. I was trying to catch my breath, but I was scared stiff. It was at least a metre taller than any of us, very broad and powerful and covered with what seemed to be a dense light grey fur. I have never seen anything like it, and I am not sure I ever want to see it again. After a few seconds the creature abruptly turned and walked away. We looked at each other for a bit, not quite sure whether we ought to*

follow or not, but curiosity won. We ran up to the hillock the creature had been standing on, but it had disappeared without a trace. We never told anyone on the base – we would have been mercilessly teased about for the rest of our stay, but we talk about it every time we meet, and we are still not quite sure what we saw."

July 14th 1984 a couple of kilometres west of Søndre Strømfjord Airport in Western Greenland. A German and a Danish tourist were out looking for musk oxen, when they suddenly saw what they at first thought was a musk oxen running directly away from them up a slight incline about a hundred metres away.

"But then the animal suddenly stopped, turned around and looked straight at us. It was definitely not a musk oxen! It was in fact what looked like a very tall and powerfully built human wearing an expensive fur-coat. It was greyish-brown, and the fur was rather long and shaggy looking, but otherwise it looked like a human being some 2½ metres tall. My camera was unfortunately still in my backpack, so I started to ease it down to the ground very slowly so as not to scare the creature, but unfortunately I lost my grip on it and dropped it to the ground. The sudden movement startled the creature, and it took off like a shot, running so fast I am certain it would be able to set several Olympic records. We lost sight of it in seconds, and though we spent the last two days of our stay in the area looking for it, we never saw it again, and as the ground was covered in plants, we couldn't find any footprints. We did find the spot where the thing had stopped, and there was an extremely unpleasant smell hanging in the air. That was all."

August 9th 1999 a couple of kilometres north of the research station on Disko Island on the Lyngmark glacier. Five Danish tourists on a walk across the glacier suddenly saw a tall humanlike creature walking across the glacier some 3 or 400 metres further up on the glacier. The creature paid absolutely no attention to them and just kept on walking.

"It was difficult to see what it was, but it was extremely tall and looked very strong and muscular."

The tourists tried to get up to where the creature had been walking to see if they could find any footprints, but they had to turn back quite quickly because of cracks in the glacier. Now what are we to make of that? I have talked to all of these people, so I am quite certain they are sincere, and that they have seen something that to them have been completely inexplicable. They were not trying to pull my leg in any way - so, what did they see?

All the sightings have a certain Sasquatch/Bigfoot like ring to them, or perhaps yeti-something ring, and had we been in Canada or anywhere else in North America, I would not have hesitated to call the sightings just that, but as far as I know, sightings of BHM's have never been recorded from Greenland before. I have never been able to locate any locals who would admit to having seen one of these things, in fact they would hardly talk about them, so I am not even certain whether they have a local name, or if they are one off's. But it seems to me we might have to add another big hairy monster to the list of those we already have.

And their man-eating habits? Well, sometimes people disappear without a trace, or sometime in olden days, old people would walk out into the wilderness at their own volition intending to die, and if their bodies were never found, well it would be natural to assume that someone or something had eaten them.

Trolls and berserkers
In the 1920's and 30's there was some talk of scientists actually having found evidence of the existence of something not quite humanlike in Greenland. During excavation at a place called Gardar Danish archaeologists found a fragment of a skull and a lower jaw so powerfully built that the individual was described as a new primitive species of human: *Homo gardarensis*.

C.C. Hansen who wrote the description, and P. Nørlund who actually found the specimen were both convinced that they had found the reason for stories about trolls and giants, and that one should be "a little careful when talking about extinct primitive humans". They compared *Homo gardarensis* to the

Neanderthals, and suggested that it might even have been more primitive than this. This individual had been found in a cemetery dating from the 12th Century, so clearly these primitive beings existed at least until fairly recently. Other individuals with similar, although not as strongly developed, traits have been found in other places in Greenland. Hansen and Nørlund even speculated that these powerfully built individuals were the reason for the stories of the much feared "berserker" warriors of the Vikings.

Within a fairly short time however, the Gardar remains were quietly shelved, as various other researchers saw them as nothing more than an unfortunate individual suffering from agromegaly, excessive bone growth, especially in the bones of the face and hands.

But the story of the Gardar Man doesn't end there – far from it. One can, if one has a tendency to root around on the internet read all kinds of conspiratorical stories about the Gardar Man, how the remnants shows a skull of enormous proportions, and that the place where the remains are stored, flatly refuses all researchers access to them, as it would only reveal, that all the scientists have been wrong for all these years, and we can't have scientists

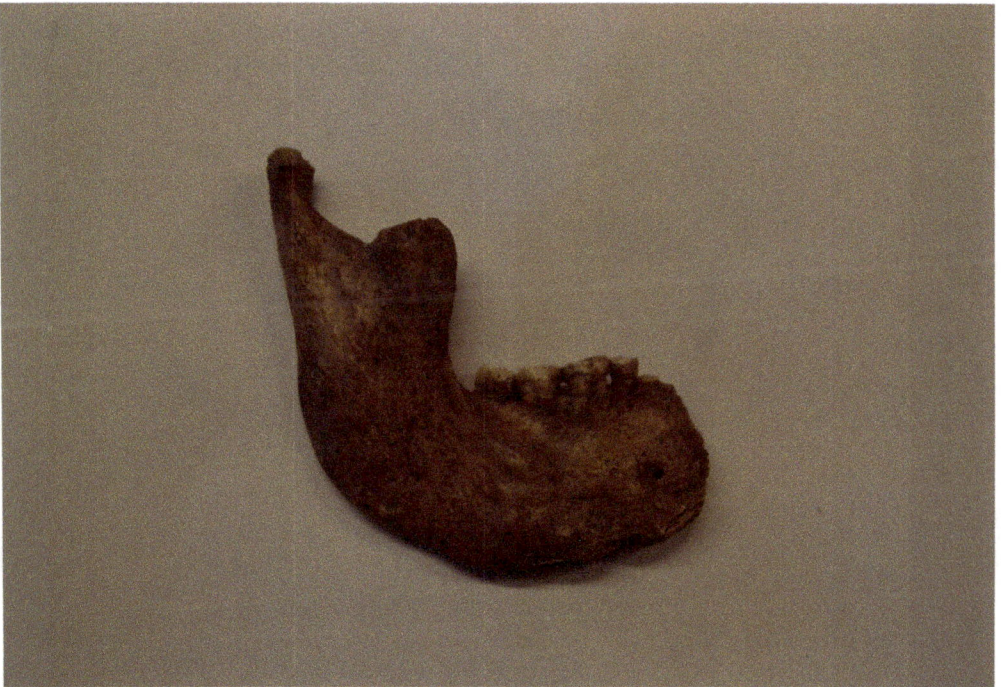

admitting to be wrong, now can we?

As the place where the Gardar Man skull and jaw is located about 10 minutes on a bike from where I live in Copenhagen, I decided a couple of years ago to test the conspiracy and see what would happened if I asked for a chance to photograph the specimens. Well, as it turned out, nothing could have been easier, the only thing they asked me to do, was to wear surgical gloves when I touched the specimens and arranged them for photography, as they would be subjected to a DNA-analysis sometime in the future.

And lo and behold, puff goes the mystery, the jaw and the partial skull was almost disappointingly small – and well within the normal size parametres of a normal human being – although with agromegaly. So no troll, no berserker and no Greenland yeti there.

The ice-jumpers
If you spend a substantial part of your waking hours studying cryptozoology and various legendary creatures, you hear and read a lot of very strange stories. Most of them pales to insignificance when compared to a story told to me by a fisherman during the summer of 1984. He had the faraway look of a man who had spent most of his life outdoors, and his face had more lines than a contour map of the Cairngorms.

We were sailing across the mouth of Disko Bay from Egedesminde to Disko Island, weaving in and out between the many strangely shaped ice-floes and icebergs coming from the giant glacier, one of the most productive in the world, at the far end of the bay. We started talking about the ice, and he told me about the dangers of sailing around it, walking on it and living in close proximity to ice every day of your life. I said something along the lines of ice being barren and devoid of life, except for the animals sitting on it, but he shook his head and started to tell me about the ice-jumpers.

I know it sound like some strange form of extreme skiing or parachute jumping, but in fact ice-jumpers are scary and dangerous ghost-like beings that live inside and below the surface of the ice in large glaciers and in the great central icecap. They don't like people very much, and are responsible for the death and disappearance of mange an unwary traveller walking across an icefield or a glacier. Before you can say icicle, a giant crack will appear, and either you will plummet downwards, to the cold embrace of the ice-jumpers, or a pair of grey transparent arms will grab your legs and drag you down to a cold and lonely death deep inside the glacier.

Occasionally you can actually see the ice-jumpers. If you sail very close to a glacier – this is actually quite dangerous, so please don't try this at home – and look up at the front of active ice, you can see the jumpers in the cracks that start to open up as the ice disintegrates and falls apart. Look closely, and you will see dark shadows jumping from side to side in the cracks, as they try to get further back inside the glacier away from the water and light that will give away their existence to the humans.

It was a fantastic story, and although I enjoyed it very much, I was quite certain, that a substantial amount of leg-pulling was going on. After all, it was only my third day in Greenland, and I just supposed I was getting the greenhorn treatment. A few days later I was talking to three local ladies in the general store in Qeqertarsuaq and told them about this story the fisherman had tried to sell me, but to my surprise, they wasn't laughing. All of them had heard the story of the ice-jumpers, and one of them was convinced she had lost her father to the mysterious denizens of the glaciers. It had happened when she was 12 years old in 1952. Her father and three friends had been out hunting, and had been walking across an arm of a glacier. For some reason they had stopped to chat for a few moments on a stretch of seemingly solid ice. For the first two or three minutes nothing happened, but suddenly a crack opened in the ice beneath her father with such blinding speed that neither he nor any of the others had time to react. Within seconds he had disappeared into the ice, and just as suddenly as the crack had opened, it closed again. Not the slightest trace of her father was ever found, and the general opinion among her friends and family was that he had fallen victim to the ice jumpers.

I was at a complete loss when it came to explaining that little story. Beings living inside the ice sounded like pure fairytale to me, and yet I met several people who claimed to have seen them as well as having lost family members to them. What to do, and what to think? Further enquiries resulted in a couple off stories more, but both so similar to the one the woman had told me that I think they were in fact all about the same case.

For a time I had to put the stories about the ice jumpers on the back burner, but that would change at the end of the summer of 1984. I had spent the entire summer in Greenland and was sailing south along the coast towards Søndre Strømfjord and the airport. The ship was part cruiseship and part passenger ferry, so plenty of time was spent on touristy things such as making several very close passes past the big glacier at Ilulissat.

Now this is one of the biggest and most active glaciers in Greenland, indeed in the northern hemisphere, so there is a lot of majestic scenery to contemplate. The ship was surrounded by icebergs in all size and shapes, there where shearwaters, gulls and guillemots all over the place, and a lonely white gyrfalcon had made a couple of passes over the boat. I was seeing amazing things, and everything was at peace, when a sound like machinegun fire signalled some serious action about to take place at the face of the glacier. Suddenly three separate parts of the glacier face seemed to be leaning out towards the ship and the water, and then started tumbling in slow motion downwards. Just as the giant chunks of ice hit the water I looked up and into one of the giant cracks that had formed in the ice. I could see at least 20 or 30 metres into the ice. And at exactly that moment some black fog-like thing jumped from one side of the crack to the other. It happened so fast, I had no time to get either my binoculars or my camera.

Surprised hardly begins to describe my state of mind at the time. Had I really seen one of the notorious ice jumpers? Unfortunately the ship was moving away from this particular part of the glacier, but we were approaching another one of the big cracks, so I readied a couple of extra witnesses, but to our great disappointment, there was nothing to see. That is until a ray of sunlight streamed into the crack, and suddenly there were ice jumpers all over the place. Now in that situation, one could of course argue, that the icy beings were trying to escape the to them presumably deadly rays, but it would in fact be more in line with reality to argue, that the ice jumpers were in fact some sort of optic illusion, probably caused by light bouncing back and forth between the faces of ice in the cracks. Combine it with a bit of wishful thinking, and perhaps the odd tragic and unexplained disappearance and voila – a race of strange and malevolent ice beings are born.

ICELAND – Fire, ice, and fairies

Iceland is a rough country. Despite its location close to the Arctic Circle, it is one of the most volcanically and geologically active places in the world. The entire island nation is placed directly across a great split in the Earth's crust that runs all through the middle of the Atlantic Ocean like a seam in a pair of trousers. Every year the entire eastern part of Iceland and the Atlantic moves a few centimetres east, and the western part moves west, making Iceland a little bit bigger every year. Consequently the Icelandic landscape is young, rough and very hard on the feet and shins if you slip. Believe me you don't want to fall flat on your face in a field of hardened lava – neither for that matter in a field of flowing lava, but that goes without saying. Iceland is filled with geysers, bubbling mud pools and active volcanoes. All that hot water by the way makes it possible for the Icelanders to grow bananas and all kinds of exotic fruits close to the Arctic Circle.

There is not much along the lines of deep dark forests, but the country is rather thinly populated (only about 320.000 people and most of these live in a couple of large cities and towns), so there is no shortage of harsh desolate landscapes, frozen glaciers, and generally scary and lonely places. Some people claim you can actually lose your mind, if you spend too much time alone in the Icelandic landscape. In actual fact Iceland were more or less completely covered in forests when the Norse people came here at first, but centuries of growing demands for timber and firewood more or less made the forests vanish completely. Today there are a few bits of original forests here and there, and in some areas people are trying to plant new forest, but it is still only a fraction of what it once was.

One would naturally assume that the Icelandic people were an equally rough lot, with battle-axes handy and swords hanging in the wardrobe next to the overcoats, but you would be surprised. The Icelandic people are one of the most well-informed and well-read in the world. Compared to their rather small numbers, the Icelanders buy, borrow, read and write more books than any other country in the world. And their interests range far and wide. One of my books, a children's book about coral reefs have even been translated into Icelandic – I kid you not! Statistically speaking one in ten of all Icelanders will publish a book sometime in their life. And Icelandic authors have won literary prices right, left and centre.

Funnily enough Icelanders also harbour a deep-seated belief in the little people and a host of other strange and mysterious beings, so there are rich pickings to be had, should you start to dig for information on weird creatures. People will often have an entire family treasury of stories about the little people, and once you get to know people, almost everyone has a sighting to report and a story to tell.

These rich picking are in a sense a mystery in themselves. I don't know if it's the water, the air or the landscape – perhaps the toxic volcanic fumes makes you hallucinate – but considering the fact that Iceland have only been populated for a 1000 years or so, the sheer volume of folklore and strange stories is quite overwhelming. At least some of the stories must have been brought over by the Vikings – especially the more gruesome and bloody ones. Just read some of their sagas, and you will see what I mean.

The landscape does play a significant role though. Until quite recently, there were very few good roads in Iceland, and like anywhere else in the world, desolate isolated areas that were hard to reach fired peoples imagination and made them tell stories – some of which may even be true.

If you look a little closer at the stories, especially their age and structure, you realize that they fall in two distinct categories. They are either old, with roots far back in time to the Vikings, or they are fairly modern – in some cases no older than the end of the Second World War. By far the most common, widespread and well known strange creatures of Iceland are the very human-like little people. There are also some large and some very large people, but they are much rarer. And then there is a selection of other weirdoes going all the way up to and including dragons.

The little people

You will be hard pressed to find a country with a greater and more vibrant belief in the little people than Iceland. Not even Ireland and their leprechauns can hold a candle to the Icelandic beliefs. In Iceland the little people are considered legitimate citizens of the country with the right to be heard in all matters political and civic. They rarely turn up in person, though 23% of the population claim to have seen them at one time or other, but according to the various stories I have collected, they have ways of letting their pleasure or displeasure be known. All over Iceland you will find roads with sudden and unexplained turns, roads going around single big rocks or trees that would have been blown to bits or cut down in any other country. To avoid these problems, the Icelandic authorities have published maps, based on the work of people skilled in talking to the little people that shows the greatest concentration of their dwellings, enabling contractors and engineers to plan accordingly. Some areas have even been set aside especially for the little people. Nobody is allowed to build or develop those areas.

One of my correspondents grew up in a little town in north western Iceland, and she has described how meetings in the local council when she started working there as a secretary, always included a 15 minutes break to allow the little people to protest against any council decisions. And the mayor always waited till the next day before signing any documents. This would give the little people ample time to protests if they wanted too. Every now and then a person who could communicate with the little people was brought in to deal with serious cases, presumably following official protests, although the exact nature of these protests were never described to her. To make matters easy for both parties, there is an academy of elf-studies in Iceland, the so called Alfaskölinn in Reykjavik where you can learn all you need to know about the little people, how to interact with them, and how to live a peaceful life with your small neighbours.

Ever since Iceland was colonized from Denmark and Norway, people have been nice to the little people. There are ancient stories about how farmers have always been careful around things like big stones where the little people live.

For many centuries, the little people and the humans were on good neighbourly terms, but when the modern civilization started coming to Iceland, problems started to arise. The belief in the little people waned, but apparently they were determined not to be forgotten nor overlooked, for modern construction projects were besieged with problems of various kinds.

In 1962, when the new harbour in Akureyri was being build, the contractors had enormous problems. Machines kept on breaking down, and even small stones that should be easily to blow up or move away with a digger, turned out to be almost impossible to dislodge. At first the mishaps were blamed on the Icelandic geology, and perhaps a little bit of bad luck, but it didn't take long before the locals started muttering about the little people and the lack of respect shown by the builders as well as the people who had decided on the new harbour. They have simply forgotten to let the little people be heard. Before this tuned into open warfare, local man Olafur Baldursson offered his service as middleman, and was able to negotiate some kind of truce. The builders had to treat the various stones and suchlike with proper respect, and then they would be moved with no fuss at all.

The whole area around Akureyri have since then been a hotbed of conflict between humans and the little people, partly because this is apparently almost the capital of the little people, and partly because this is a very active area of Iceland with a lot of roads connecting industrial areas, urban areas and so forth. If you drive around here today, you will find quite a number of roads with unexpected bends and turns around big stones and boulders – dwellings of the little people. The road to Kopavogur even have a turn that bypasses one of the little people main castles.

But by and large, things have a tendency to work out all right in the end. Every now and then you get a modern contractor or possibly a firm from abroad that refuses to pay the little people the proper respect, and then the trouble starts all over again. During road construction in northern Iceland in September 1995, a bulldozer kept breaking down every time the construction crew tried to remove an enormous boulder that blocked their way. Geologists suggested the boulder was in fact much larger than it appeared, and that they would need heavier machinery to move it. The company didn't want to spend a lot of money bringing in more expensive equipment just for the sake of one boulder, so instead they tried a local wise woman, who got in touch with the little people, who told her, that it was all right to move the stone, if it was done in a decent and respectful way. The builder complied with the instructions the lady forwarded from the little people, and the stone could now be moved with no trouble at all. It even turned out to be considerably smaller than everyone had expected.

You can see similar traces all over Iceland, and should you get inspired to go exploring, the area around Eyjafördur also has a considerable population of little people.

Redheads
A personal meeting with one of the little people must be extremely exciting, and in almost every single case it is described as something very positive, something that in many cases have been actually life-changing for the human part of the meeting.

The little people have been described in various ways, but if we try to make a general description it would be something like the following. As their name would imply, the little people are in fact small, but maybe not as small as one would expect. They are somewhere between 1 and 1,5 metre in height, they have red hair and hairy faces, they eat flour and grain, and they live in big stones, big mounds and so on. With a size like this one would imagine they should be quite easy to see, but they are quite crafty and good a keeping hidden, no doubt helped quite a lot by the fact that they can also change their shape and size. But for some strange reason, although they were generally wary of humans, they were less wary of children, so the vast majority of sightings of the little people were done by children.

When I was at school, my now sadly late teacher of Danish, who was born and raised in Iceland, used to tell stories about going to school when she was 7 or 8 years old with small groups of little people as spectators along the way. They never talked to her, and she never talked to them, but she could always see them out of the corner of her eye, nudging each other and pointing to her. She ascribed their interest to the fact, that she was fairly small for her age, and had flaming red hair, although it faded considerably when she grew older. She moved to Denmark when she was about 14, and lived the rest of her life there,

but she did go back regularly, to meet with her family, and the little people, but when she was about 18 or 19, the little people disappeared, and she never saw them again.

It was exactly the same for one of her friends, who usually met the little people when he went out fishing. He preferred to fish on his own, and he would always have company, usually within 5 or 10 minutes after he'd started the actual fishing, he would look to the side, and there would be a little man sitting on a rock or in the grass, just looking at his lure and generally minding his own business. They never talked, but he always got the impression that the little man was enjoying the company. Should anybody else pass by, the little man would put a finger to his lips, smile – and disappear in the blink of an eye. This boy stayed in Island for the rest of his life, but just like my teacher, he lost contact with the little people when he came out of puberty.

In a strange way, this reminds me of the imaginary friends some young children have – and even more so in the next story which concerns a couple of twin daughters that grew up in Seydisfjördur in eastern Iceland in the 1960's. I interwied their mother about ten years ago, and she told me how the girls would drive her crazy with their stories about playing with the little people and talking to them in a very special language that only they could understand, but which they absolutely refused to demonstrate for their friends of parents. A couple of times the girls stayed out all night – at this time they were about 10 years old – and couldn't for the life of them see why their parents were making such a fuss, when they returned home next morning. They had been with their friends, and they had of course looked after them very well. 3 years ago I contacted the two daughters and tried to get them to tell me about their experiences with the little people, but they absolutely refused to discuss them apart from saying that they both felt that what the little people had taught them were the main reason for the two of them having success in various artistic endeavours – one of them is a highly successful musician, and the other is a potter whose works are in great demand.

The unfriendly ones
By and large the relationship between the ordinary people and the little people is amicable and friendly. It has to be – although the population of Iceland is fairly small, there are apparently enough humans, and more than enough little people for them to come into contact on such a regular basis they simply have to be on good terms unless life on Iceland should be in a state of constant warfare. Some elf-scholars reckon there are at least twice as many little people in Iceland as there are humans.

Things are not so good when it comes to the larger creatures, which also inhabits the Icelandic wilderness. Those creatures are much rarer, and that is probably a good thing, as they are far more malevolent and dangerous. They have absolutely no qualms about making a damn nuisance of themselves, wreaking peoples stuff, or even abducting young and luscious maidens – or perhaps juicy maidens is the prober way of describing this, as several of the various trolls and giants are in fact cannibals and like nothing more than to feast on a good thigh. Some of them have even more interesting habits – female trolls sometimes abduct strong young men - not for eating mind you, but to be used as lovers. Yes, even trolls have need and feelings.

But – and this is quite important. The distance between humans and trolls are not as great as what one might expect. You can actually make friends with a troll, and a better and more loyal friend you will not find anywhere. So if you need a deadly and dangerous enemy, or a fantastic and trusted friend – look no further than …

The rather large people

The creatures in this group are a strange and varied lot. They are the size of humans or larger – sometimes quite a bit larger – and they have a lot of things in common with creatures like bigfoot, almas, yetis i.e. they are big, hairy and smelly, and they run around on two legs. But they also have things in common with ghouls, zombies and ghosts. And that makes them rather dangerous.

Icelandic ghosts f.inst are by and large a nasty lot. They are more zombie-like, rising from their graves and stomping around at night, especially for some strange reasons on the roofs of people's houses. Sometimes the noise is deafening, and sometimes it is only a soft scraping sound, so I can't help but wonder whether it is in fact inspired by the sound of birds or perhaps arctic foxes or something like that frolicking on the rooftops. I know from personal experience that owls and martens can make an awful racket, if they decide to use your roof or attic as a stomping ground, and I use the word stomping quite deliberately.

Anyhow – whatever you do, and no matter how much noise there is emanating from your roof, and how little sleep you are getting because of that, you have to stay in bed, because you do not, and I repeat, you do not want to meet an Icelandic ghost up close, because if you do, the drangr or aptrgrangr, as they are known in Icelandic, will quite simply break every bone in your body, and that is guaranteed to kill you, or at the very least destroy the rest of your evening. The bonebreaking feat is probably caused by the fact that Icelandic ghosts have superhuman strength and several magical abilities to boot. The usually live in old Viking graves – perhaps they are in fact the ghosts of Viking warriors – they can change shape and size and they smell distinctly iffy. Just like skunk apes and other big hairy monsters from other parts of the world. They also have the ability to drive livestock mad, or at least scare them out of their senses, and that is an element well-known from other parts of the world where the presence of a bigfoot or something similar is often signalled by scared dogs, terrified horses or similar happenings. Oh – and they can also swim through solid rock by the way.

Beware of trolls

By far the most common species of rather large people in Iceland are trolls. They are a deeply rooted part of Nordic folklore, and they are, or at least used to be, quite common in Iceland, but the number of sightings is quite low. The last officially recorded sighting I know of is from the 1750's, but through my various contacts, I have been able to unearth a few later ones. And to give you an idea of what kind of sightings we are dealing with, here's three samples:

May 27th 1967 just outside Keflavik Airport.

Petur, who at the time was a 27 year old carpenter was on his way to work. He was putting of various shelves and cupboards in a house fairly close to the airport. It was fairly early in the day, just starting to get light, and quite misty and cold when he started hearing heavy footfalls a little bit behind and to his left on an open field. Because of the mist he couldn't really see anything, but a sudden gust of wind thinned the grey curtain a little bit: " and suddenly I saw a very large man standing in the field maybe 50 metres away. He didn't say or do anything, he was just standing there looking at me. I couldn't see him clearly, but he was either very hairy or wearing a very big fur coat. I looked for other people, but there was nobody out, so I tried yelling "Who are you?" "What do you want?" But the big man didn't react. I didn't know what to do, but then I tried to run towards him. But as soon as I started to move, he turned and disappeared into the mist. I never saw him again."

July 9th 1981 of the tip of Langanes in northeastern Iceland

The crew of a fishing wessel passing the Langanes peninsula going North,were adamant they saw two extremely large humanlike figures walking slowly along the coastline, apparently gathering driftwood. The two figures were extremely large, as witnessed by the fact that one of them picked up an entire tree trunk and just stuck it under "his" arm. The captain of the fishing trawler absolutely refused to sail any closer, saying that getting close up and personal with a couple of trolls would only end in

tears.

January 29th 1988 on Highway 1 south of Skaftafell

A truckdriver going west along Highway 1 was passing close by Skaftafell Mountain when he saw what he described as a very large man running away from the road and up towards the top of Skaftafell. He was apparently moving so fast that he disappeared from sight within seconds.

All those trolls

As can be seen from these three descriptions and numerous others by the way, trolls are by and large humanlike, but it is also quite clear, that we must be dealing with several different types or species if you like. Trolls are humanlike indeed – although in some cases rather hairy, but not always, they can actually be so humanlike it is difficult to tell the difference. And they definitely don't have horns and tails like you see in fairy-tales and children's books. They are extremely strong though, and can be quite a bit larger than normal humans, but they are not giants, although some form of trolls can be quite odd-looking, especially when they grow old and start to develop extra heads or limbs. Very old trolls can have three separate heads or four or five arms. And trolls can get quite old – several hundred years according to some stories.

Despite their rather rough look and a tendency to have a very volatile temper, trolls are in facts extremely intelligent, not the dimwitted morons form we know from fairytales. They have a vast knowledge of things, including information that is not usually available to humans. So if you can make friends with a troll, the road to success is open. And although trolls are extremely vary of humans under normal circumstances, they are also extremely loyal. On the other hand they can also make formidable enemies capable of carrying a grudge for far to many years should you accidently rub them up the wrong way.

There are also several stories about trolls helping humans to catch monsters or other kinds of weird and dangerous creatures, although that might be modernized versions of the story about how the Nordic god Thor tries to catch Jormungandr, the Midgaardworm.

4 trolls

If you go through the various troll stories and pick out the things describing the way the trolls look, you end up with 4 different kinds of trolls:

- Your average everyday common troll looks more or less like a large human, although a bit on the hairy and muscular side.
- Night trolls look more or less the same, but they are, as the name implies, nocturnal, and usually a bit more hairy than the day trolls, and of course they have the rather strange weakness of turning into stone when hit by sunlight, although that may be a very specialized form of camouflage enabling them to hide in plain sight.
- Some trolls are considerably larger than others and have an unfortunate tendency to eat humans, especially young and juicy ones. The stories about this type is few and far between and do sound more like some form of bogeyman stories designed to keep the children away from dangerous places.
- Finally we have the volcanic trolls. They are only found around and in the various volcanoes in Iceland. They are very big and squat and rough around the edges. According to some stories they are actually made of stone or lava, and have a tendency to set fire to any thing flammable they happen to come into contact with.

But of course, the big question is not how many trolls that live in Iceland, but whether we actually have any kind of physical evidence for the existence of these interesting beasties.

Big bones

One grey and dismal summer's day sometime in the 18th century a gravedigger in the village of Illugastadir was in for the fright of a lifetime. He was busy digging a grave for one of the inhabitants of the little town who were to be buried the next day, when he came across a bone. Now he was quite used to stray bones turning up in the graves he were digging, but this bone was something out of the ordinary. It was the thighbone of a man, and it was enormous. The poor gravedigger was terrified – he dropped his pick and shovel and ran to get the priest.

At first the priest found it hard to believe the story the gravedigger was telling, but on the other hand he could see the man was genuinely frightened, so he agreed to come back to the cemetery with him, where lo and behold, he could see and touch the

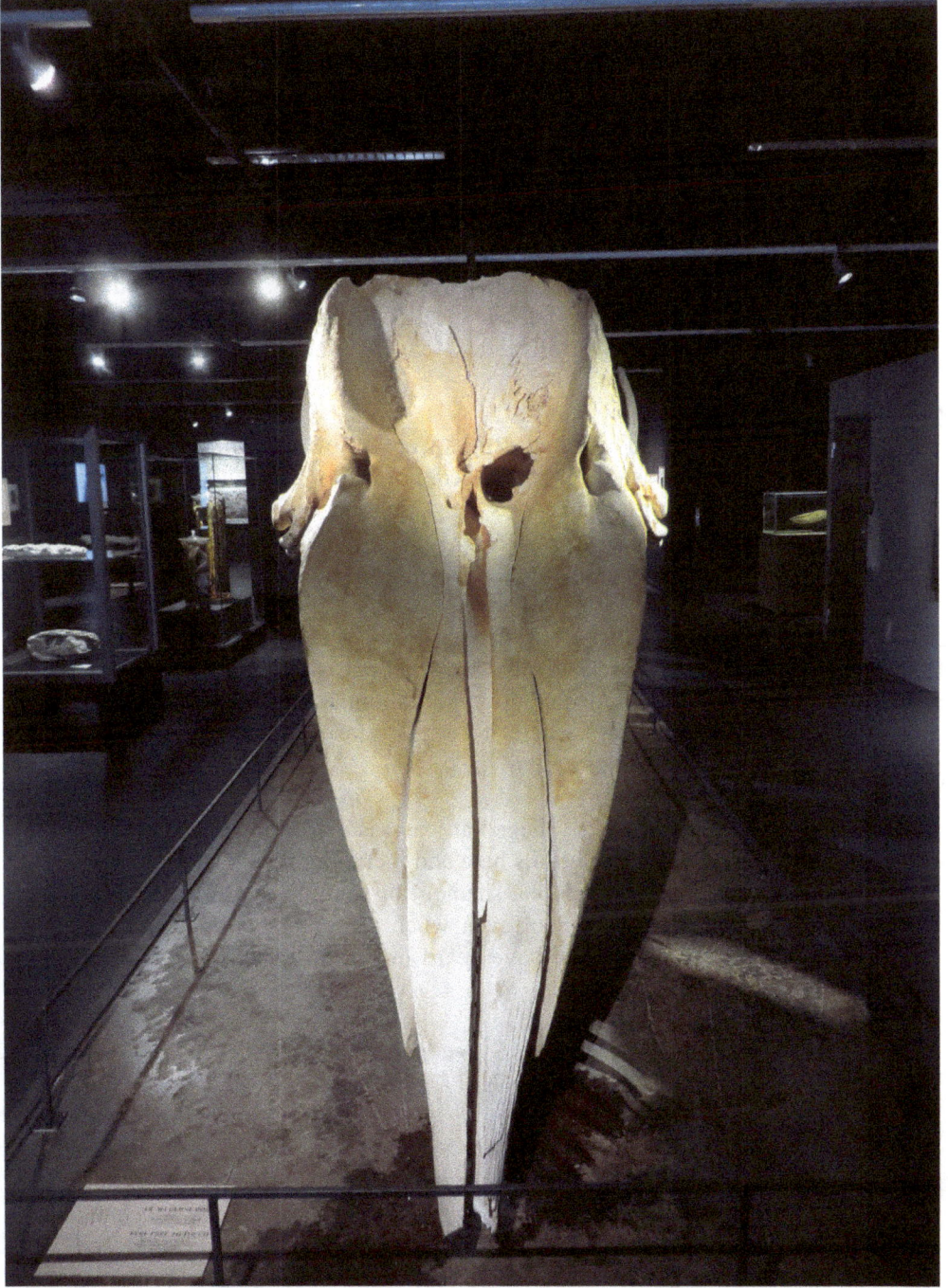

enormous bone for himself. Now Illugastadir was not a big place, it still isn't, so within a very short time everybody knew what had been found. No mere mortal could grow to that size, but then some of the old folks remembered a story they had heard many years before about a man who had been taken by the trolls. Everybody thought they would never see him again, and that the trolls would fatten him and eat him, but somehow he managed to escape and came back to his village. He thought he had been gone for only a few days, but the trolls had kept him for months. They had treated him like an honoured guest, had given him plenty to eat, but he wasn't allowed to leave the trolls cave. First he thought he would end up in the trolls cooking pot, but apparently one of the female trolls had taken a fancy to him, and wanted him for her husband. He had tried to fight off her advances, but had no hope of being able to do so for ever, so one day when the trolls left the cave, he ran off as fast as he could back towards the village.

He ran like a man possessed certain the trolls were following him, and finally reached Illugastadir where people were very surprised, not to say frightened to see him again. Not only because they thought he was dead, but because he had grown – a lot! Apparently troll cooking makes you grow. He was now twice the size of a normal man. Unfortunately the escape had taken his toll, so he died a couple of days after telling his story.

Unfortunately the priest thought all this was the work of the Devil himself, so the bone was hastily reburied, and nobody talked about it anymore. No drawing were made of it, and nobody knows where it lays. A closer look of his remains would probably yield some interesting results.

Actual physical remains of trolls (alleged at least) is extremely rare. There are a few stories about people finding dead trolls here and there, but for some reasons the bones of the trolls are never kept, which is a shame. For some reason the Icelandic people haven't really bothered about it. Not even when something fairly major happened. Just take a look at this story that I have heard in several different versions over the years:

This is something that happened a very long time ago. The story was told to me by my grandfather, and he had heard it from his grandfather, and perhaps the story is even older. But the story goes like this. One morning some people were walking along the beach at Raudhasandur about 300 km northwest of Reykjavik. They were gathering driftwood for their fires, when they saw something very big lying at the far end of the beach. At first they thought it was a dead whale, but when they got closer, it looked like nothing they had ever seen before. The animal or whatever it was, had the size of 10 men. It was hairy all over, had two arms and two legs, but they were mangled and broken, and parts of them were missing. The head was also gone, and all that was left was a long thin neck. They ran in terror back to their village and called upon an old man to come and see the thing. The man was very old, and took a long time to walk to the beach, but he at once said that they had found the remains of a troll, and that they should give push it back out to the sea as quick as possible. Later they heard that the troll had come from Norway, and had tried to wade all the way. It had been sent by the king of Norway to check on the people of Iceland, who at that time apparently had been showing signs of unrest. The troll was certain it could manage the trip to Iceland, but apparently something happened on the way, and it was killed.

If this story has any basis in fact, what those people found were probably some very dead and decomposed marine animal. The alleged fur and four broken and mangled limbs, and the neck without the head makes me think of a dead basking shark or another big shark although I am a bit confused as to why people wouldn't have recognized that. Surely a people so at home at sea would know how dead and decomposed whales, seals and so forth would look – but perhaps not. Also it seems rather strange that the troll would have been walking all the way from Norway – where it was found were about as far away from Norway as you could possibly get. Why not go ashore on the east coast and then walk across the land?
Why not stop these relevant asides and get on with it? Ok, ok!

Faces in rocks
Another place in Iceland where you can find evidence for the existence of trolls is Dimmuborgir near the famous Lake Myvatn, where seven lava pillars are supposed to be trolls caught out by the daylight and turned into stone. These pillars can at least go some way in explaining the stories about trolls turning into stone (or into trees as they do in Scandinavia). Every now and then you come across a rock formation or a tree that looks distinctly human. It is perhaps a bit rough around the edges, the features are not exactly delicate, but they are so lifelike you are almost forced to conclude, or at least you were, if you were living in the Viking ages or the Middle Ages, that this rock, tree or whatever had once been alive. Should one be of a sceptical nature, one might even suggest that this is how all the stories of troll started. If the trees or rocks were once alive, what had these creatures been doing when they were alive, how had they lived, and could they perhaps explain some of the natural phenomenon and mysterious disappearances that had everybody perplexed and scared.

And so – with the creation of the trolls, suddenly we have an explanation for those fuzzy figures sometimes seen on fogbanks, or on low-lying clouds, or the optical illusions that make people seem far closer than they actually are of much bigger, or perhaps even smaller than they really are.

Horses and other forms of magical livestock

The Icelandic people are extremely proud of their horses. Their exact ancestry is a little bit hazy, but they are descended from the first horses the Vikings brought with them to Iceland from Scandinavia. According to the Medieval Icelandic manuscript Landnamsbogen (The Settlement Book) the first person to settle in Iceland was the Norwegian chieftain Ingólfur Arnason in 874 A.D. He was followed by other Norse chieftains of Scottish, Norwegian and Irish descent, who sailed to Iceland in their small open boats with their families and the best of their livestock. This was how the first horses came to Iceland, and since then there has been virtually no genetic change. At the end of the 13th century Iceland came under Norwegian control, and that was the start of several centuries of almost total isolation of the Iceland, and an extremely limited contact with other countries. And during all that time, the Icelandic horse kept plodding along the rocky plains of the island.

The Icelandic horse is a short, stocky and extremely tough and powerful little horse. It can be found all over the island and are as well known to the Icelandic people as their own children. So when the Icelanders sometimes claim that they see horses of another kind, it is worth paying attention. Not the least because the Icelanders for many years have had a very strict rules regarding import and export of horses. If a horse leaves Iceland, it is never allowed back in, so you can be quite certain that the only kind of horse on Iceland is the Icelandic horse. So – what of the other ones?

Stories of strange horses can be found in most of northern Europe. The horses are often extremely odd looking in some way. They may look like they are just about ready for the knackers yard, or their feet may be on back-to-front, or their legs are swollen. In fact these horses are often extremely strong and/or fast, or they have some rather interesting abilities, such as being able to change size of shape or swim and dive in streams, lakes or indeed the sea. Some of them even dives of high cliffs when spooked by humans and disappears into the sea. Most of them are fairly peaceful and benign. There are stories about people getting lost in the volcanic wastelands and then meeting up with a strange horse that gently guides them towards a road or a house.

Some of the horses though, have are rather more malevolent disposition, and you are well advised to keep clear of them. It is especially important that you refrain from trying to ride them. If so, you might suddenly find yourself as glued to their backs and unable to get off when they either run out into the water, which will inevitably result in you drowning miserably, or they will take you on a run of such breathtaking speed, and then throw you off, that you either break your neck or end up being not quite right in the head afterwards.

The horses of the little people

In all these cases people explain the look and behaviour of these rather odd horses with the fact that they belong to the little people, mer-people or other non-human beings. Even trolls (the smaller ones) have horses. They are often immensely strong – and a bit on the big side, but not overtly so. The Icelandic versions of these horses that sometimes go under the name of hykor, are not as weird as some. They are a bit skinny and scrawny compared to the stocky animals you see and ride on a daily basis in Iceland, but that's about it. Apart that is, from some rather interesting leg problems.

The following story was told to me by an Icelandic doctor I corresponded with for a while in the early 1990's:

> *This story has been told in my family for almost a hundred years, so I hope you find it interesting. When my great-great grandfather was a young man in the late 1800's, he was walking home along the beach late one night. He was living north of Reykjavik with his family in a little house quite close to the sea. It was a cold night, but there was a full moon out, so he could see fairly well. He was hurrying home, hoping his wife had some late supper waiting for him, when he saw a horse a bit further on walking slowly along the water's edge. He was quite close to home by now, so he got worried that maybe one of his own horses or perhaps one of his neighbour's horses had escaped. The horses didn't pay any attention to him, so he walked quietly closer in an attempt to perhaps grab its reins and lead it back home. But when he was a few metres away, he realized that this was no ordinary horse, in fact he was quite certain he had never seen anything quite like it. The horse was walking along with its head hanging so low it was almost dragging on the sand. He couldn't make out too many details, but the animal looked sick and thin, and it walked slowly and gingerly. Its left leg also seemed swollen, and the horse was clearly reluctant to put too much weight on it. It was altogether a sorry sight, but even so, it looked wrong somehow. It took my grandfather a few minutes to realize something so strange, that his nerve suddenly broke, and he ran home like the devil was on his heels. The horse's hooves were back to front! It was not a horse, it was a hykor!*

If we forget the hooves for at minute, there is nothing supernatural or mysterious in this story – it is a very sick and tired horse plodding along the sand, nothing more, nothing less. But those back to front feet are a bit difficult to explain. It is of course

not uncommon for horses living wild to have split edges on their hooves or various minor splits and cracks just like carnivores can have broken claws and humans bad nails, and I can actually well imagine, that every now and then you could come across a horse with hooves so badly damaged that the front would look like the back, where there actually is a deep cut in the hooves. If you just saw it in poor light, you probably would think that its feet were actually on backwards. Split hooves would probably also mean tender feet and explain why these horses are walking gingerly and putting their feet down very carefully. I am though at a loss to explain the swelling of the foreleg (and for some reason it is nearly always the left). I have spoken to several veterinarians to see whether there is some form of disease that might explain it, but none of them could think of something reasonable.

A possible explanation for the existence of these weird horses is some degree of inbreeding. The current world population of Icelandic horses is close to 200.000 and as a rule they are extremely healthy, but they were isolated for a very long period from the 13th century onwards, and in the late 1700's they had to pass through a very tight evolutionary bottleneck, when Iceland was hit by Möðuharðindin (The Trials of the Fog) from 1783 to 1785 following the eruption of the Laki volcano. A fifth of the human population died, and three quarters of the horses, leaving only 6000 animals left alive. Every single Icelandic hose on the planet today are descendants of them. So does some form a genetic malfunction manifest itself in the form of a strangely shaped horse with abnormal feet?

Scaly things

The official Icelandic list of reptiles and amphibian species is very short, incredibly short in fact – actually non-existing. There are not a single species of snake, lizard, frog or toad living on the island. The Icelandic climate, especially in winter is far too long and harsh for any member of those animal groups to survive. Unofficially though, the story is quite different. Sightings have occurred of various small, not so small and some very big reptiles.

The furry snake

In the summer of 1983 something rather strange was afoot or on the loose in Heiðmörk about 10 km southeast of the center of Reykjavik. This is a very popular recreational area, and it has been designated a natural reserve since the 1950's. There is a fair amount of trees growing here, and more than 150 species of wild plants have been recorded for the area. In July and August that year, at least 4 different people, claimed to have seen what could only be described as a furry segmented snake.

> **Vigdis, 32:** *I was sitting on a rock painting a watercolour of the landscape. It was a very quiet day, so I could hear all kinds of small noises, birds squeaking, people talking somewhere of in the distance and a car going by somewhere. Suddenly I also heard a strange rustling noise coming from the ground just in front of my easel. I couldn't really see what it was, but when I stood up I realized it was a snake. I was very surprised. I had never seen a snake in real life before. It was small, only about 30 or 40 cm long, and it was dark, grayish brown. It was slithering along quite quickly, and it made me think of a toy train set my brother used to play with. The snake moved like it was segmented, like it was made of little parts. The strangest thing is, I think it had fur. I know snakes should have scales, but it looked like fur. And then suddenly it was gone, and I never saw it again.*

As Vigdis so rightly put it – snakes should have scales, so what on earth did she see that day? A furry segmented snake in a country with no snakes? It is of course conceivable that somebody had lost a pet snake, and that it somehow it had found its way to Heiðmörk, but what of the fur and the apparent segmentation? It doesn't really sound like any kind of snake known to man anywhere, let alone in Iceland. Which is why I think the explanation could possibly be found amongst the rather sparse mammal fauna of Iceland.

It is a little known but well documented phenomenon, that shrew families sometime play train, when they are out and about. The first young bite the tail or the behind of the mother, and then the next one bites his or her sibling and so on, until a long chain of shrews has been formed. They then proceed to go out like this looking for food. I have seen this a couple of times in Denmark, and apart from the fact that it looks faintly ridiculous, it does in fact makes you think along the lines of a small furry snake. The only problem with that theory is of course that there are no shrews in Iceland. But there are mice, wood mouse and house mouse to be specific, both species brought inadvertently to the island by humans. As far as I know, the train phenomenon has not been described for mice, but why shouldn't they? It makes it possible for the family to go about its business in a fairly efficient way without losing anybody on the way. Wood mouse are a bit on the big side, and I suspect it would have been possible to see the individual mice in a chain, but house mice are smaller, and they are often darkish grey or grayish brown. So a family of mouse out and about in a rather unusual way for a single summer? I have never heard about the furry snake before or since 1983, so that's my theory until something better comes along.

The rocky iguana
There has also been a single sighting of something the eyewitness described as something akin to a large iguana like the land iguanas on the Galapagos Islands or perhaps the big rhinoceros iguana. This sighting was made in the summer of 2002 by Brad, an American student visiting Iceland during his summer holidays. I was told the story in 2009 when it turned out that this student was actually a friend of one of my friends. I met him by chance at a party at my friends place, and we ended up talking about animals, cryptozoological and otherwise, and this was when he told me this little story:

"In 2002 I was a geology student at university. My great-grandfather came from Iceland, so when summer came, I decided to spend my vacation there, visiting relatives and perhaps look a little bit at the geology of Iceland. I decided to spend a few days walking around the Jökulságljúfur National Park. They have a lot of interesting volcanic rocks, and at that time I was hoping to become a volcanologist. One day late afternoon I was walking at the foot of a cliffside with a lot of loose stones and piles of broken rocks. I just happened to look up the slope and got the chock of a lifetime. About thirty feet up the slope a very large lizard was sunning itself on the rocks. It was extremely well camouflaged. It has the exact same colour as the rocks, but I could clearly see the legs and the tail, and I also saw a slight movement of the head. This was totally unexpected. As far as I knew, no such thing could be found in Iceland. I decided to investigate. I stood there

looking at it for some time. It didn't move at all, and seemed to take no notice of me, so I started climbing the slope to get closer. It was a bit rough going, and for at time I had to keep my eyes on my hands and feet. That was a bit of a shame, because when I got up to about where I though the animal was lying, it was gone. I sat there for a while and scanned the rocks for it, but I couldn't find it anywhere. I have no idea what it was, but it looked like one of the big things from the Galapagos."

Having spent several hours talking to Brad I am quite certain he is not lying, and that he is convinced he saw this animal. But I am afraid I think his senses have deceived him. The fact that the animal was completely motionless and had the exact same colour as the rocks makes me think that what he saw was in fact rocks, that because of a certain play of light and shadow, simply looked like a big lizard, and then Brads imagination filled in whatever gaps there might have been. This would also explain why he couldn't find any trace of it, when he got closer, and the angle of the sunlight had changed and turned the rocks back into rocks. The slight movement of the head he thought he saw, was probably something he imagined as well.

Scary scaly things
Apart from these odd singular sightings of what seems to be perfectly ordinary reptiles, but a very long way from home, there are also sightings of something much much bigger; because some people claim to have seen dragons flying about their draconian business over the rugged landscapes of Iceland. The stories are few and far between, and not especially detailed, but I am including them here for the sake of completeness. Now Iceland was o course populated by Vikings, and they had a very special and close relationship with dragons. The used them as symbols of power and strength to strike fear in the hearts of their enemies. They put them on the bow of their ships, on their shields and sometimes also on their clothing and other items.

As one could readily imagine, the Icelandic dragons have a special affinity with the countries volcanoes. According to some stories they actually seem to be living in them, and one must concede that using the crater of an active volcano as your front door is a pretty good way to keep unwanted guests out and away from your treasure – or whatever it is dragons keep in there.

And then of course there is the highly dangerous skoffin – but more of this a bit later on...

And so there are in fact a few stories of people claiming to see something very large and dark either diving into one of the volcanic craters or flying up from it. The details are usually fuzzy, not the least because people have trouble seeing what's going on with the heat, steam and smoke from the volcano obscuring their vision. Most of these stories are fairly old, but I have one from as late as 2010 when the volcano Eyjafjallajökull erupted and grounded planes all over the northern hemisphere because of the enormous amount of ash it had thrown up into the atmosphere.

After a couple of days of activity a local man went out to have a look at the enormous column of ash and smoke, and it was while watching this, from several kilometres away, he swears he saw something very large circling in the ash and smoke.

At first he had what he called a Superman moment – is it a plane, is it a bird...no it's. Yes, that's the question, what is it? Realizing at what a large distance he was seeing this, the flying thing must have been enormous. It seemed to have to very large wings, but apart from that it was just an enormous "blob" moving around the ask and smoke. In an e-mail to me he wrote:

> *"As I see it, there are only two possible explanations, either I was imagining the whole thing, or what I saw was in fact a dragon. I find that very hard to believe, but then again I don't think I could imagine something like that."*

Apart from the volcanic sightings there have also been a couple of people claiming to see something very large flying over at tremendous speed at tremendous height. No details were given, it was just something large and dark and moving very fast.

I know this isn't much to go by, and I am sorry to say I don't actually think there is any kind of biological explanation for any of this – geological perhaps, or even meteorological, but that's about it. The dark shapes leaving and entering craters, as well as the giant flying thing seen during the Eyjafjallajökull eruption was probably nothing more than flying clouds of ask, smoke or steam taking on a vague dragonlike shape for a few seconds, letting the imagination of the eyewitnesses do the rest. As for the high-flying dragons? Clouds maybe, or perhaps nothing more special than birds. The cold dry arctic air can sometimes make things look vastly different from what they really look like.

This little mention of birds leads us nicely towards the next group of strange beings in Iceland, the birds- but before we get there, we have to return to something I mentioned above, something extremely dangerous, and strangely enough a creature that starts of as a bird, but then turns into something far more sinister.

The skoffin

The skoffin is a strange and terrible being. Luckily it is also rare, because it is extremely dangerous. It is also the reason most Icelandic chicken runs only have young cocks in them, because on very rare occasions old cocks will lay an egg, often small and malformed, and from that egg a skoffin will spring. At first it looks like a tiny worm, indeed sometimes you will actually find tiny worms in eggs when you crack them, especially from older hens.

I have seen one of those worms myself at the tender age of 8, when I was helping an aunt who lived on a tiny smallholding about 75 km west of Copenhagen. I had helped her gather the eggs from the henhouse, and now we were about to bake a cake. I was cracking eggs in a bowl, when a tiny threadlike completely white worm suddenly emerged from the mass of yolks and whites in the bottom of the bowl. I got a bit of a chock, but my aunt just picked it up and threw it in the rubbish. And then we kept on baking, but I must admit I was a bit apprehensive when the cake was served later that day. I still remember it quite clearly and I made several drawings of the egg and the worm at the time. I still got a couple of them.

Luckily my aunt knew how to deal with the tiny worm, because it the worm gets a chance to find a dark place to hide, a well or a deep cellar or something like that,

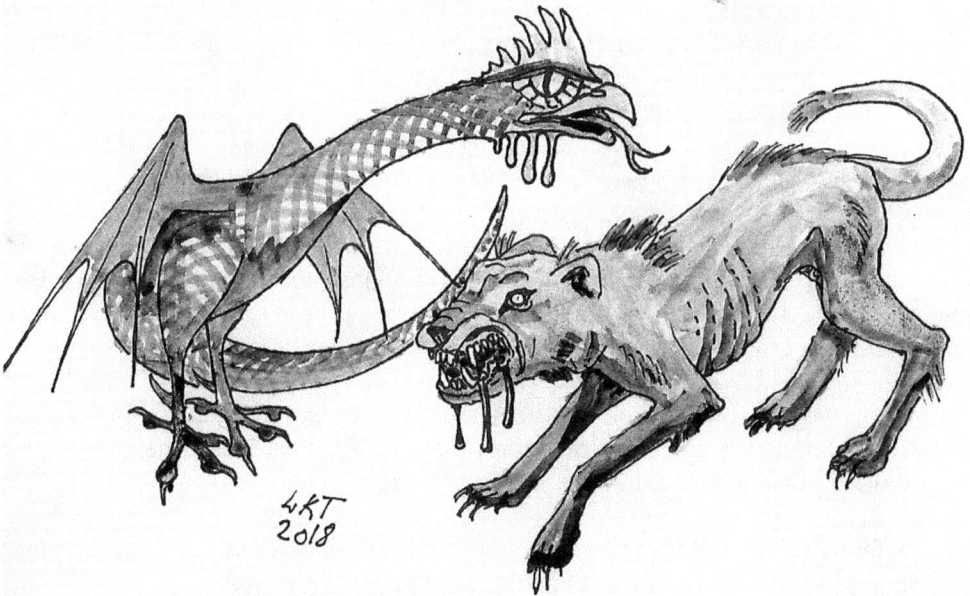

within a very short time it will grow to a terrifyingly large size. It will turn into a large snakelike dragon, or a dragonlike snake if you will, often with a beak and a large comb on its head. According to some legends it will also have a pair of wings, although they can be quite scrawny and apparently not much good when it comes to flying. The worm has now turned into a skoffin. So now it's a good time to put as much distance between the skoffin and yourself as you possibly can. Skoffins, or basilisks, as they are known in other parts of the world, are extremely dangerous. They can kill you just by looking at you, or perhaps just by their mere presence. There are of course no eyewitnesses of fully grown basilisks, but quite a number of sightings of strange worms in chicken runs, as well and strangely shaped eggs and weird looking cocks.

Compared to European basilisks, skoffins are almost peaceful. They prefer to stay away from people and find somewhere dark and dismal to hide far out in the Icelandic wilderness. But of course, should a mountaineer, somebody just out for a walk or perhaps a hunter come across the lair of skoffin, it is very unlikely they will survive – some of them may never be seen again. Indeed one of my Icelandic correspondents, who grew up in eastern Iceland in the years between the two world wars, told me that in his childhood it was a common expression when somebody disappeared, that he had been taken by a skoffin. And of course every now and then people would be found dead somewhere, with no signs of physical damage, perhaps with a look of astonishment on their faces, and then of course you would know, that the last thing they had seen was a skoffin. As late as 1996 I heard that local was saying that a man found dead along a road in northern Iceland, had been killed by a skoffin.

Before you start altering your plans for your next holiday to Iceland or get nervous when a weird looking egg turns up among your own chickens, it is in fact not uncommon that older hens, especially in small private runs, where they are often allowed to live considerably longer than in a commercial venture, develop various kinds of hormonal imbalances. They can start to develop combs and wattles, and end up looking like perhaps rater smallish and short-legged cocks, and the standard of their egg-production starts to fluctuate. Their eggs can get smaller of rather inconsistent shape-wise. Sometimes they even produce shell-less eggs.

And the tiny white worms are in fact nematode worms, extremely common internal parasites in all kinds of animals. Sometimes they can even get inside the shell of the eggs, so when you crack an egg – hey presto, a young skoffin.

But – before you think all is safe, I am afraid there is another skoffin, and to some extremt, it is even more dangerous than the first one.

Dogs of the underworld

To understand the nature of this other skoffin, which incidentally also has the ability to kill people with a single glance, we first have to go back in time a little bit, in fact all the way to the time of the old Norse gods, where Frigg, the wife of Odin, drove around in a carriage drawn by dogs. Consequently dogs were of extreme importance to the Vikings. They were revered as hunting companions, guardians and companions, but also as symbols of the goddess, and being on friendly terms with Frigg was quite important. Among other things she assisted with childbirth. So dogs were of this world, but definitely also of the world of the Gods, the afterlife and so forth, meaning that they also had supernatural powers.

One of the most important duties the dogs had to fulfill was the guiding of the souls of dead people to the underworld. Normally these guard dogs of the dead were by and large benevolent beings making sure that the souls did not get lost on the way, and were forced to wander as a ghost until Ragnarok, the end of the world. But things can go wrong. There is always some trickster around to throw a spanner in the works.

The other skoffin

> ***Summer of 1986:*** *A man driving along Highway 1, the Icelandic coast road, close to Djupivogur, saw what he described as a small brownish, bald and decrepit looking dog shuffling along next to the road. When the man slowed down to have a closer look, the animal turned its head and look at him. He was so terrified he nearly crashed his car, but*

managed to get it under control and stop it. Not only was blood dripping from the jaws of the animal, its eyes were also flaming red. But the worst was still to come. As he was sitting there, petrified with fright, the animal gave took a couple of steps towards his car, snapping its jaws, and then it turned and walked away. And he swears that it walked across a wide water-filled ditch – on the water's surface!

Winter of 1981: A woman living in Reykjavik had gone to the cemetery at Fossvogur to visit the grave of her husband. It was getting dark when she got there, and there was quite a lot of snow on the ground, but she got a terrible fright, when she approached her husband's grave. An animal was sitting on top of the tombstone, looking at her with glaring eyes. "It looked like a cross between a dog and a very large cat. It was whiter than the snow, and although it was small, it looked so angry I felt like a mouse about to be eaten. It felt so evil I still shudder when I think about it. Suddenly it gave out a snarl, jumped off the stone and disappeared. And I swear it did not leave a single mark or footprint!"

If you were to ask an elderly Icelander about these two creatures, there would probably only be one answer - skoffin. This kind of skoffin is the result of a probably ungodly union between an arctic fox and a cat, and every time an arctic fox is involved, you can be sure, that things will go completely sideways, as they are sneaky, crafty and devilish creatures, not entirely of this world. But we shall hear more of this wily thing shortly. First we have to deal with its nightmarish spawn.

These scary things are either looking like rather demonic cats with eyes so bright they can burn a hole in your soul or perhaps they manifest themselves as strangely thin, emaciated and almost bald scrawny excuses for a dog. But don't allow yourself to be fooled for even a second. Like all other kinds of skoffin, they are at best only dangerous, at worst absolutely lethal. They are considered death omens in some parts of Iceland, but they are also quite capable of killing people on their own, if the poor people haven't already keeled over from pure fright. The cat-like skoffin is also known as the ghoul cat, and according to legend, it has a rather nasty habit of breaking into cemeteries, graves and mausoleums and eating the dead.

The Yule cat
And then of course, there is the jólaköttur, or Yule cat. It is by far the weirdest and most dangerous of Iceland's mysterious cats. This thing knocks the alien big cats of other countries into a cocked hat. As the name implies, it only shows its viscious countenance around Christmas time. It belongs to one of the Icelandic giants. His name is Grylle, and the Yule cat is in fact his pet. Normally he keeps it under tight

control, but around Christmas time, he gives it a bit of a run, and then you have to run as well, because the Yule cat eats people, especially if they haven't received any new clothes before Christmas!

The wily werefox

And so we have to return to the arctic fox for a short while, partly because I think it is responsible for most, if not all of the sightings of skoffins and ghoul cats, and partly because it has been revered and feared in Norse folklore for centuries. One has to remember that the arctic fox was the only land mammal when the Vikings arrived in Iceland, and that must have made a huge impression, especially when they Norse people saw how easily the fox handled the rather brutal wintertime.

The arctic fox became associated with the trickster god Loki very quickly, probably because of its fantastic ability as a burglar and mischief maker. I have personally experienced arctic foxes break into a tent with people inside and steal their butter.

Later on the fox developed a rather more sinister reputation – it was supposed to have shape-shifting abilities, it could be dangerous in various ways, and as arctic foxes are also great scavengers, it got a reputation of eating the dead - which in many cases it probably did. So even in the early 1900's you could meet people who firmly believed you could only kill an arctic fox if you used a silver bullet or a weapon made of silver.

Of giant eagles and clever ravens

Birds are a very visible part of life in Iceland. The country has a small but varied fauna of birds with several interesting species from all white gyrfalcons to colourful harlequin ducks. And because of the country's geographical location in the middle of the North Atlantic, it also get's it's fair share of rare birds from North America as well as Europe and in some cases even Africa.

Birds have always played an important role in the everyday life of the Icelandic people. They have hunted them in their thousands, especially ducks and auks, and in one sad case even killed the last known examples of an entire species, when the last definitively known specimens of the Great Auk was killed on a little island south of Iceland in 1844 by three Icelandic fishermen. But it is only fair that birds should be important in Iceland. After all, one of the four guardian spirits of Iceland is a bird – a giant eagle. This comes from an ancient legend about a wizard sent out by a Danish king to check out Iceland and its inhabitants. He transformed himself into a whale and swam north. When he got to Iceland he met four mighty creatures – the guardian of the south, a large rock giant; the guardian of the west, an enormous bull; the guardian of the east, a mighty dragon with an entourage of smaller reptiles; and finally the guardian of the north, an eagle so large its wings were touching the sides of the mountains when it flew through a valley.

With credentials like that, one would imagine a huge amount of folklore about strange birds, but that is not in fact the case. There are some stories, mainly about eagles and ravens, and I have managed to dig up a few more, but that's about it.

The stormriders

Nobody can be in any doubt that eagles are most impressive birds. Some of them look like flying barndoors armed with something along the lines of armour-piercing talons the size of steak-knives, and most of them has a beak that looks fit to bit a tin can in half, or at the very least tear the roof of a medium-sized car. Oh, and they have almost armour-piercing eyes as well. I have seen my fair share of white-tailed sea-eagles, and I am quite sure, that if looks could kill, I would have been reduced to a small pile of cinders many years ago. As it is, I feel a distinct urge to apologize profusely and get lost every time an eagle nails me to the spot with those diamond hard yellow eyes.

But to be perfectly honest – eagles are not THAT big. A lot of it is actually the sheer force of the bird's personality. So how come you can find people claiming to have seen absolutely giant birds, eagles presumably, but not necessarily, flying over at enormous heights, or far out at sea? And these birds are also claimed to ride along on stormfronts, or to actually bring the bad weather, or at least serve as an omen of bad weather (thunderbirds anyone?).

I am reminded at this point of the old Arabian tales (Sindbad and all that) about the enormous roc bird capable of flying with an elephant in each set of talons and one held in its beak. Quite a number of stories about this creatures tells about seeing it flying past far out at sea. Some authors have suggested that people have actually seen storm-clouds or perhaps mirages of distant mountaintops. This may very well be the case in Iceland as well, but in this part of the world, there is another natural phenomenon, that might explain some of the sightings of the giant eagles, or storm-birds, or whatever they are.

Arctic waters are the home of enormous breeding colonies of the little auk – and that goes for Iceland as well, although you have to go very far north to find the really large colonies. These birds are only the size of starlings, and just like starlings, they sometimes gather in enormous flocks. I wonder if some of the sightings of stormbirds are in fact distant flocks of little auks? Their movement would look very much alive – but from a great distance, you wouldn't be able to discern the individual birds, all you would see would be a big dark "thing" moving along at great speed. Just a thought!

The evil one

I think you would be hard pressed to find a bird with a more sinister reputation than the raven. It has been a symbol of evil and the powers of darkness for centuries. Authors and film-makers have used the big black birds to great effect, and sinister poems have been written about them – if you haven't read it already, I sincerely suggest you find a copy of Edgar Allan Poe's "The Raven" somewhere, and read it – Nevermore!

According to legend, ravens are inherently evil and vicious, and they have general assemblies twice a year where they decide who is going to die during the next six months, and who is going to survive. But you can in fact make friends with them if you remember to feed them regularly. They will still eventually be the portents of your death, but not before it actually is your time to go.

From a cryptozoological point of view, the most interesting thing is in fact a handful sightings of ravens with various big white splotches. This may not sound like much, just a couple of leucistic or albinistic birds, but there might be more to it than that. But, as the same type of raven was once fairly common on the Faroe Islands, they will be treated in some detail in that chapter.

The red beetle

To end the chapter on Iceland and its mysteries on a weird note, here is a little entomological mystery told to me by an Icelandic student, living in Copenhagen about the same time I was putting the finishing touches to my finishing thesis at the University of Copenhagen. His name was Askur, and he was from somewhere in the north of Iceland. He never actually told me, what he was studying in Copenhagen, but I met him on a regular basis at the August Krogh Institute of Physiology, and we spent many afternoons and evenings talking about a diverse range of subjects. Askur had a vast knowledge of weird stuff, and one day following a long talk and a substantial amount of alcoholic beverages, he told me the strange story of the red beetles, and if you are reminded of Edgar Allan Poe's famous story The Gold Bug, not to worry, so was I.

It all started during the summer of 1976 – a summer that brought strange happenings and weird creatures to many parts of the world, but that's quite another story. Late one afternoon, Askur was sitting in his room in his parents' house, when he heard a strange knocking sound on the window. He tried to ignore it at first, but finally got up to investigate. And lo and behold – there was a big beetle slowly making its way across the windowsill. Askur was rather surprised with his find, because not only was the beetle quite big, 3-4 cm in length, it was also bright red. Now the insect fauna of Iceland is fairly limited compared to say northwestern Europe, so although not an

entomologist in any way, Askur did know all the major groups, and he was quite sure, he had never seen anything like it before. It looked like more or less like a ground beetle, although it was a bit on the heavy and powerful side, but the bright colour was extremely unusual. And what was even more unusual was the fact that there was more than one. Just as Askur was about to go and find a jar or something to catch the beetle in, he realized, that the entire outside wall of his parent' house was literally crawling with the things. There were hundreds of them.

But he wasn't the only witness to this strange invasion. A large number of gulls had discovered this instant buffet, and was descending on the neighborhood by their hundreds. In the mean time, the first beetle had disappeared, and now it suddenly seemed to be a race against time and the gulls, to actually catch a specimen, so Askur dashed down to his parents kitchen to try and find something to catch one of the animals in, and then went out to actually do so, but he literally had to fight his way through hordes of hungry gulls. Alas – he never caught one. The gulls ate them all, or drove them away, and although Askur spent several days scouring the neighborhood and talking to other people, he never saw another of the red beetles, and even stranger, he never found another person who had seen them.

I can only surmise that perhaps southerly winds had brought the beetles up from perhaps North America, or maybe even Africa. Askur told me that he had actually spent quite a lot of time trying to find a match for the beetles in various books about North American and African insects, but he never could. Anybody know anything about big red beetles, feel free to comment!

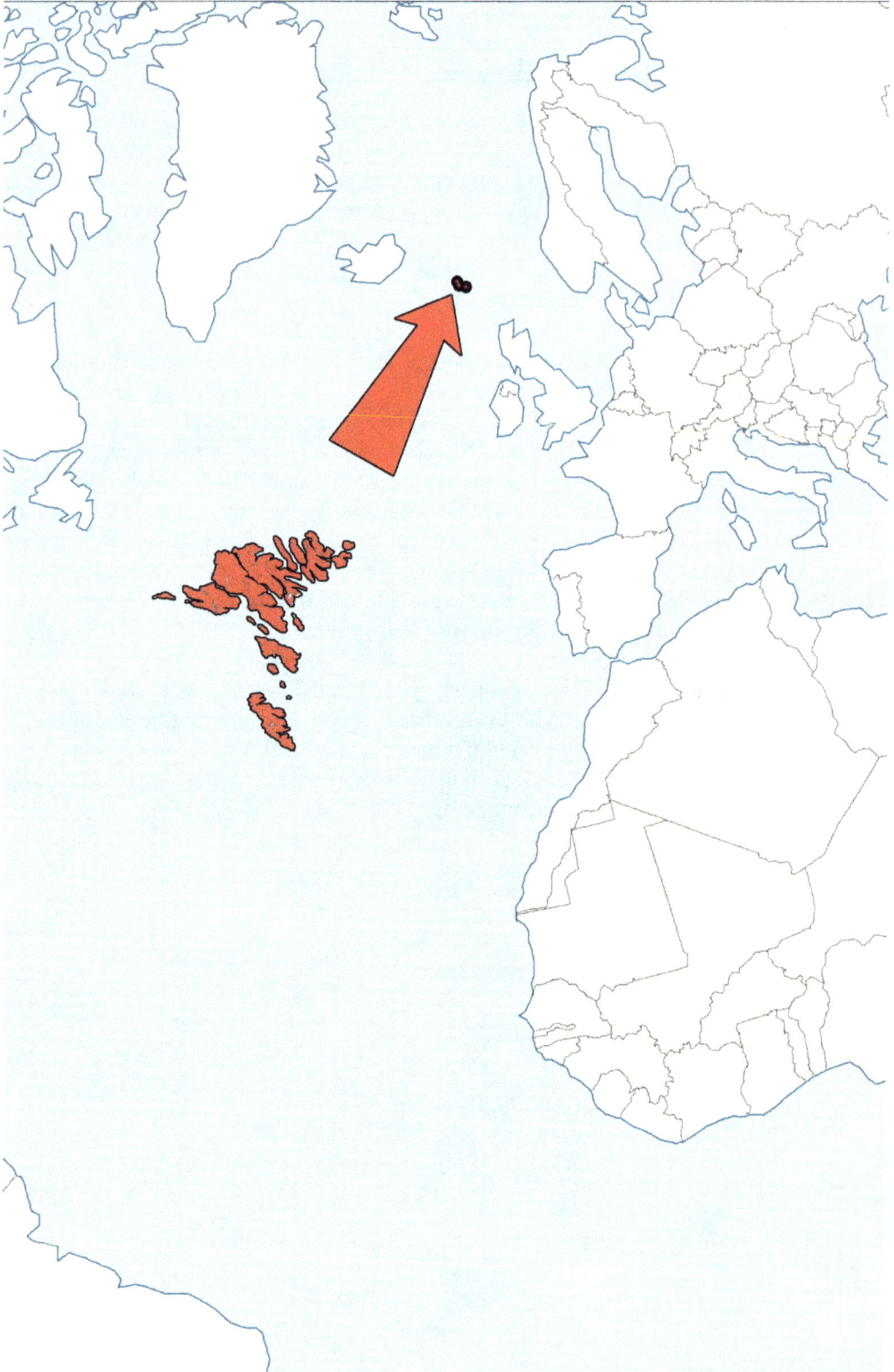

THE FAROE ISLANDS - 18 small green islands

he Faroes is a tiny group of island located in the North Atlantic roughly halfway between Iceland and Norway. It is a small place with only 18 islands and something in the neighbourhood of 45,000 people. There is not much in the way of plant or animal life – probably because it rains a lot, and the amount of sunshine is low, and the islands are a long way from everywhere. The three dominant animal groups are humans, sheep and birds. This is not a place with room for much in the way of strange beings – they would have a devil of a time hiding themselves, should they be any bigger than your average squirrel. There are no forests to speak of and the remote and isolated mountain ranges, if such a word can actually be used, are never more than a few miles away. Nevertheless - a word I find myself using to an increasing degree in this book - there are a few stories about strange, weird, wonderful, and in a couple of cases extremely scary creatures afoot on the islands.

You can for instance see some rather strange birds on the Faroes, and I don't mean just from a twitcher's point of view, but also birds that basically shouldn't exist.

In recent years sightings of strange beings have become increasingly scarce, maybe because the creatures themselves are becoming rarer, or because the people of the Faroes have been busy involving themselves in financial scandals or trying to fend off the increasing amount of criticism coming from international animal welfare groups because of their continued killing of pilot whales, and consequently haven't had time to keep an eye out for weird animals. But never fear – a few weird ones have managed to establish themselves, and we are going to take a look at them – although one of them may not be anything more than a genetic aberration of an already well known bird.

Winter's tales

In a place like the Faroe Islands, there probably was not much to do in winter before the invention of radio, TV, and suchlike electronic devices. Instead people had to resort to the making and telling of terrifying stories about scary and dangerous creatures of the night. Some of these stories have a clear functional value in keeping kids and other reckless individuals from going walkabout on dark and stormy nights, and people of a perhaps morally-lax constitution from straying too far from the straight and narrow, as it were, although they may sometimes have been used as an excuse for various digressions. According to a guy from the Faroes who was at the university at the same time as me, quite a lot of his fellow Faroese people were quite fond of straying from the straight and narrow, and of making up tall tales, so he advised caution when analysing stories and eye-witness accounts from the Faroes.

Legends like these have been told – and are being told – all over the world. What makes the Faroese versions a little bit different is the fact that people have claimed, and are in fact still claiming, to meet these scary creatures on a regular, albeit rare basis.

The Gryla

The Gryla is a terrifying nocturnal creature actually known in various from all over the world. It is one of several species of monsters that punish children and/or adults that haven't behaved or done something they were supposed to do, usually things of a religious nature. On the Faroe Islands it is a monster carrying a sack in which to stuff children that haven't behaved, or more specifically haven't observed the fast between Lent and Easter – a custom well known from Catholic countries, but rather strange in a small group of islands in the middle of the North Atlantic.

In a series of stamps issued by the Faroe Postal Authorities, the gryla has been depicted as a tall, bluish and very skinny humanlike figure carrying a sack, a big knife and with a long row of featherlike projections along its back.

This is of course an artistic interpretation of the gryla – there a quite a lot of stories about it, and even some actual sightings, and they vary quite a bit, although there are some similarities, as the following sightings will demonstrate. Although I do suspect, that gryla is in fact a bit of an all including term meaning simply monster, just like the Icelandic skrimsl or skoffin.

Sometime in the early 1920's *the grandfather of J.D. who now lives in*

Copenhagen, and one of his friends, were living in Klaksvik, the second largest town on the Faroe Islands. Late one evening in October they were walking home from a birthday party in one part of town to their home at the other end of town. On their way home they passed very close to the coast, when they saw, what they at first thought was a very large piece of driftwood. It looked like an entire tree bent over in a strange angle. As they approached, they discussed whether to try and drag the tree further up on dry land, as driftwood was a valuable commodity. But when they were about 50 metres from the tree, it suddenly straightened itself and stood upright. It was not a tree, but a very tall, almost skeletal two-legged being. The two friends froze in their tracks, as it looked like the creature was scanning the horizon, looking for something. They both had grown up with stories of the gryla, and they were in no doubt, that this was what they were looking at. It didn't have a sack or a knife, but the creature was radiating an almost physical sensation of irritation and menace. The two friends had dropped flat on the ground, when the tall being turned and started walking away along the coastline with long rhythmical strides. They were far too scared to follow, and instead opted for running home as fast as they could. It took them several weeks to gather enough courage to

tell their families about their strange meeting. The reception was decidedly mixed, and within a very short time, everybody simply stopped talking about it. J.D. only learnt about it, when he helped clear out his grandfathers things, after he had died in the 1970'es, and found a small diary where he had described the encounter.

I have seen this diary, and I actually think the story had been written the same night, because the handwriting is so irregular and shaking, that some passages are almost impossible to read. I have no idea what the two friends had seen, I am inclined to put it down to a mixture of overactive imaginations, perhaps a drink or two (or three or four...), but it must have seemed very real, as it had clearly scared J.D.'s grandfather almost out of his ability to write.

In July 1951 a fisherman, the father of an old friend of mine from the University of Copenhagen, was approaching Nolsoy in his boat, when he chanced to look upwards toward the top of the island beyond Nolsoy Lighthouse. Now Nolsoy is covered in grass, and under normal circumstances there is not a tree in sight anywhere. But quite suddenly a tree seemed to have sprung up on the very top of the island. As my friends father was staring open-mouthed, the "tree" suddenly keeled over, and revealed itself to be some kind of skinny four-legged thing, that made him think of pictures of dinosaurs he had seen in books. It had apparently been rearing up on its hind legs, because now it was standing on all fours, and sporting a strange row of spikes or long featherlike projections along its back. At this stage my friend's father had to attend to his boat for a few seconds, and when he looked up, the thing had disappeared. Most members of his family were of the opinion he had been seeing things, but according to my friend, his father insisted to his dying day, that he had seen a living moving creature.

December 6th 1981 a 14 year-old boy woke up in his room late at night. There was no noise of anything else that could have disturbed him, but nevertheless he felt distinctly like something was wrong. He tried sniffing for smoke, thinking that something had caught fire, but everything was very quiet. When I interviewed him 20 years later, almost to the date in early December 2001, he explained that the strange stillness scared him more than anything else. For a few minutes he debated with himself whether to wake up his parents or try to go back to sleep, and decided to try to get back to sleep. But just as he was about to close his eyes again, a strange bluish glow seemed to seep through the curtains and battle through the darkness of his room. Very nervously he got out of bed, walked over to the window, lifted the edge of the curtain slightly and peered out into the night. What he saw scared him so much that he ran screaming through the house and into his parent's bedroom. According to a letter I later received from his mother, he was completely beside himself with fear, and it took her almost an hour to get him to tell her that he had seen a tall skinny man walking past the house with a sack on his back. The figure had been giving off a strange blue glow, but it didn't do anything apart from walking past the house with long almost bouncing strides. With some difficulty, the parents managed to convince their son that he had only been dreaming, and finally persuaded him to go back to bed, but although the fear left him, he is to this day still convinced, that he saw a real being.

I for one, am inclined to think it was some form of lucid dream, apart from one tiny little fact niggling at the back of my mind. According to my correspondent one of his friends living a bit further up the street claimed to have seen the same figure...

Just for the record, I do have some experience with lucid dreams – not myself mind you - but one of my oldest friends from university. She was a lucid dreamer if ever there was one. Her dreams could continue for seconds, sometimes minutes after she had woken up. Among the dreams she has told me about was the time when she woke up to find her bedroom completely covered in cobwebs - they all slowly melted away, as she was lying in bed looking at them. And when they had completely gone, she just got up, and started her day. On another occasion, the room was filled with roses that did the same disappearing trick. Lucid dreams can be very real, but as far as my friend goes, she had never seen any actual figures.

And so the mare...

And then of course, there is the mare – and no, it is not a female horse, it is something much worse. Actually the mare is known from all over the world in various forms. The mare in responsible for nightmares, or as we say in the Nordic countries "mareridt", literally meaning ridden by a mare, some sort of demon that straddles the chest of the sleeping person and causes terrible terrifying dreams. Mares can take on various shapes – an old shrivelled and wrinkly woman, a small furry demon, or as is the case in the Faroes, a tall, thin, almost skeletal grey being, without any distinct features, mouth, ears or nostrils, but with a pair of large black eyes – greys, anyone?

Anyone with access to UFO-literature will be able to read about hundreds if not thousands of sightings of grey beings with big black eyes turning up in peoples bedrooms. I have no idea if there is any connection between 'the greys' and the Faroese mare, but it is kind of interesting that some stories from the Faroes tells about

the mare taking people to other worlds in their dreams, and sometimes torturing them. I am not familiar with ufo-sightings from the Faroes, so I don't know if the grey have put in an appearance there as well, but whatever the circumstances, I have never been able to find a, shall we say, independent sighting of a mare. All the stories that I know off is about people lying in bed unable to move and seeing the mare sitting on them, and then disappearing, as they wake up from at nightmare. An interesting phenomenon, well deserving off a closer scrutiny, but in my opinion not cryptozoological, so I will not treat it any further in this book.

The screaming babies

A common form of superstition on the Faroe Islands is the belief that an illegitimate baby, killed shortly after birth, will return to haunt just about everybody - especially whoever killed it. Legends like this are known from many parts of the world, and so is the fact that the hauntings will continue until the dead baby gets a name. Apparently things are not good in the afterlife, should you end up there in a nameless state.

The haunting can be physical, i.e. the baby returns as a tiny, sometimes furry creature, in some cases no bigger than a ball of yarn, rolling around on the floor, on the ground

or sometimes even in a church during the sermon, singing sad songs about how its life had ended even before it had gotten its name. Some of these ghostly babies are known as screamers, as they wail and moan and complain screaming their misery out across the Faroese landscape, sometimes by day, but mostly by night. In many cases you can't actually see them, but you can hear their ghostly wailings, and you can in some cases be totally mesmerized by them. Some of the screaming babies have almost siren-like attributes, being capable of luring you to your death, by getting you to walk of the tops of high cliffs, or simply walk out into the sea and disappear.

In olden days, the screaming babies usually showed themselves as themselves so to speak, but as we approach modern times, the physical manifestation of small strange looking babies becomes rarer and rarer until they seem to disappear completely sometime in the late 1800's. Or at least I haven't been able to find any stories later than 1876. But the sound lingers on. I have dozens of stories from people who have been out and about, usually at night, where they have heard the wailing and screaming of lost babies. In many cases they have spent hours searching for the lost baby or lost child, but have never been able to find them. It is usually impossible to pinpoint the source. Sometimes it is as if the sound is coming from the sea itself, sometimes it comes from deep within the ground, or even from somewhere up above in the darkness of the night sky.

I have talked to several Faroese people about this, a couple of modern university students, a doctor and a painter, and have suggested, that there might be a perfectly natural explanation for these strange nightly wailings, which usually brings the conversation to a halt. Some of them are indignant of my suggestion, that a people who have lived in close contact with nature for centuries, should not be capable of recognizing the sounds of the animals that surround them. You see, I think that the wailing babies, or rather the sound of the wailing babies, is in fact the sound of young seals (in Denmark we actually call them wailers or howlers), or the sound of shearwaters or similar sea-birds, going about their business deep within their nesting tunnels, chattering, groaning , wailing and grunting. If you haven't heard the racket a colony of shearwaters or storm petrels can make, you can't possibly imagine the complete otherworldliness of their sounds. They are probably responsible for the sound in the air as well, as they can be active at night. And the sound of a baby-seal wailing its heart out is absolutely heartbreaking.

Some people take issue with that explanation for completely different reasons. They want to believe in something supernatural, something unexplainable, and keep on insisting that science doesn't know everything, which of course is entirely true, but still... And knowledge about the animals of your surroundings, perhaps because you hunt them for food, does not necessarily mean you know their sounds, their behaviour or what they do in their spare time. I have experienced exactly the same phenomenon, when I some years ago suggested that most if not all of the sightings of the monster in Lake Storsjön in Sweden were in fact of swimming moose. The locals did not take kindly to that idea, and wondered whether I actually dared to suggest they did not know what a moose looked like. As a matter of fact that was precisely what I did, so conversation became a bit strained from that point on.

No matter how much you think you know about the animals in a particular area, you will be surprised by how much you don't know. Stand at a coastline or a beach at night, and I can almost guarantee you will see or hear things that are new and strange and leaves you baffled, or perhaps even scared. You might even see a troll!

The beachtroll and its kin
In olden times there was a widespread belief in trolls in the Faroes, especially beach trolls or trolls living near the top of the various rather steep cliffs of the islands. They are indeed daunting, and should you fall, you will in some cases have plenty of time to see not only your whole life pass before your eyes, but several reruns and special sections in slow-motion. It is dangerous, and you have to be careful, so it is no

wonder, that most students of folklore thinks that stories like this are a special creation, a form of safety equipment if you will, designed to keep mainly children away from dangerous areas, especially at night. And who knows, perhaps the odd adult as well.

It is an excellent theory, apart from one single little fact. There are people, reasonable, sane, well-educated people, who claim to this day, that they have actually seen the trolls, or whatever they are, because when you look closer, they are not as troll-like as the proper trolls you find in for instance Norway or Denmark, they look much more humanlike, they walk only on two legs, they are quite hairy, and in direct contrast to what the old stories advice you to do, they have in some cases been credited with saving lives, although in most cases, they seem to ignore the terrified people who sometimes see them.

Oh yes, and I almost forgot – they are by no means alone. The Faroe Islands are the home of a host of human-like beings –considering the small size of the place, it is a miracle they don't bump into each other constantly. The general name for all of these creatures are the "huldre"-people, and that name covers beings as diverse as trolls, elves, brownies, dwarfs and several others. How the fit them in is truly beyond me.

***Summer of 1993:** A good friend of mine decided to spend part of his summer holidays on the Faroe Islands. Today he is a leading expert in Northatlantic fishes, but at that time he was fresh out of university, and wanted a little break, time to do some birdwatching, and perhaps rummage around the fishmarket in Thorshavn to find fish he hadn't seen before. He had a good time, despite the fickle weather of the islands. One day he had gone for a long walk on the island of Streymoy. He thought he had prepared himself thoroughly, but as it turned out-he hadn't. After a couple of hours walk it started raining in a way that made him think of the tropics. He couldn't see a thing – the rain was coming down so violently his glasses was completely misted over, and if he took of the glasses, he would be more than likely to step over a cliff without noticing. He was looking desperately for a place to find some shelter, when a tall figure suddenly rose from the grass about 50 metres in front of him, and beckoned him closer. He thought he had stumbled into another hiker, but it soon transpired, that this was a hiker like no other. First of all he simply couldn't catch up with the other one, no matter how hard he tried. He tried yelling at him, but got no reaction whatsoever. But as he was well and truly lost, he kept stumbling after the tall figure. As time passed he realized two things, that his guide seemed to be dressed in a heavy fur coat, and that he seemed to have* *bare feet. Despite the rain, my friend was certain he saw the naked sole of a foot, when the other hiker stepped over a stone. And at one point, a sudden gust of wind blew from his "guide" towards him, and assaulted him with a nosewrenching stink of unwashed animal. After stumbling around like this for about an hour, my friend suddenly realized he was standing on a road, and that a car was coming towards him. He looked around for his saviour, but he or perhaps it, was gone as quickly and silently as it had appeared.*

Autumn of 1999: *P.P. who lives in Aarhus in Denmark, had received an invitation to a wedding in the Faroe Islands. One of his work colleagues, who were of Faroese descent, were to be married in his hometown. P.P. was not looking forward to this. When I interviewed him, he came over as an extremely restless and fidgety person, and he is not especially interested in nature and beautiful scenery, far from it. The more shops, bars, cafes and buildings, the better, as far as he is concerned. For various reasons P.P. arrived for the wedding two days early. His friends were busy with preparations, so he was very much left to himself, and within hours he started to get bored, so to kill a few hours, he decided to go for a long walk. This by the way, all took place on the island of Eysturoy close to the town of Fuglefjord (literally the Fiord of the Birds). P.P. informed his host, that he was going for a walk, and asked for suggestions as to which way he should go. His host, an elderly relative of his friend, who had agreed to house several of the wedding guests, made several suggestions, but advised him to stay well clear of a particular area. He wouldn't tell him why, but insisted he would be sorry, if he went that way. So P.P. naturally set out in precisely the direction he was advised not to. "I thought it was all superstitious mumbo-jumbo, so I set out just to prove the old man wrong. The first couple of kilometres went well, and I was actually enjoying the view, but suddenly I started feeling cold and ill at ease, although it was a nice sunny day for once. I started thinking about friends and relatives I had lost, and within minutes I was close to tears. This was extremely unusual for me, so I tried to pull myself together, but somehow that only made it worse. And then a few moments later, I was hit with what I can only describe as a something akin to the beam of light from a lighthouse, but it was a beam of extreme anger and hostility. Someone of something very much wanted me to go away. By now I was beginning to feel seriously scared, and decided to head back as quick as I could, but then I froze in my tracks. About 100 metres in front of me, a tall figure in a fur coat was standing with its arms outstretched, not in a* *welcoming gesture, but clearly signalling "no further". At that moment my mind went blank. I must have turned and started to walk back, but for the life of me I can't remember doing it. When I came to my senses, I was within sight of the house of my host. And he was actually standing outside the door, waiting for me. As I came closer, he handed me a glass with a very stiff drink, motioned for me to empty the glass, and then quietly said – I told you so..."*

So what do we make of these stories? They are, I might add, just two out of about twenty similar sightings that I have been able to collect. I am perfectly willing to consider the existence of unknown primates or something similar in areas like Siberia or the Himalayas, where there are enormous areas of empty wilderness for them to hide in, but that simply doesn't exists in the Faroes. So, if the sightings cannot be explained by straightforward lies, overactive imaginations, or miss-interpreted sightings of other people or perhaps sheep out and about, the tall hairy two-legged beings must be some form of manifestation, a zooform phenomenon, as cryptozoologist and head of CFZ Jonathan Downes have called these animallike "beings" that seem to be not quite as physically solid, as one would prefer.

The next animals we are going to take a look at though, are very physical indeed, although probably extinct by now. At the Natural History Museum in Denmark I have personally handled and photographed no less than 6 of...

The black birds
Or actually the birds that are not as black as they should be. Any twitcher will tell you that isolated islands are magnets when it comes to rare birds. You can see the most incredible species in places like the Faroes, the Orkneys, the Shetland Isles and various similar places dotted all over the North Atlantic. Usually it is just one or two lone, tired and sometimes rather dishevelled individuals that turn up, but one upon a time, the Faroes were actually the home of a rather strange form of bird, the like of which the world has never seen, and probably never will again, although the odd alleged sighting is reported every now and then.

The black sinister looking raven with its deep croaking voice have always had a very special place in Nordic folklore. After all Odin, the chief god, had two ravens, Hugin and Munin ("Thought" and "Memory"), that served as his spies, telling him everything that went on in the world. That of course tended to make the common man somewhat suspicious of the big black birds. The ravens' habit of eating the bodies of the dead on the battlefield, often starting with the eyes, did not exactly endear them to the populace at large, although people had a grudging respect for the birds' intelligence and cunning. On the Faroe Islands ravens weren't exactly popular. For

many years there was a law requiring people to kill a certain amount of ravens every year, as they were considered vermin capable of killing lambs and possibly even babies.

But the ravens of the Faroe Islands were, at least for a time, not like the other ravens of the north. In amongst the black birds, which in their heyday could be seen in flocks of several hundred birds, you could also find birds that were partially white.

You do occasionally find animals with a few white feathers in amongst their normal coloured one. It is fairly common in f.inst. blackbirds, but the pied ravens of the Faroes were something else entirely. At time they were fairly common, and although no two birds looked the same, the general pattern of their colouring was the same. For a time scientists even entertained the idea, that the pied birds were a species in their own right, but today everybody agrees, that it was a special colour morph of the particular subspecies of raven found in the Faroes and in Iceland. Strangely enough pied ravens have never conclusively been shown to have existed in Iceland. Although a few people claims to have seen partially white ravens there.

The first writings on the white-speckled ravens are pre-1500's, so they have been

Pied raven. By Ascanius (1723-1803) who was a Norwegian biologist and mineralogist and Fellow of the Royal Society who studied under Carl Linnaeus.

Føroyskt: Hvítravnur (Corvus corax variu)

Føroyar
400
HVÍTRAVNUR Corvus corax varius

known for many centuries. As early as 1655 Danish naturalist and scientist Ole Worm describes two specimens he had procured for his Museum Wormianum in Copenhagen sometime before 1650. Until about 1850, the standard coloured raven was very common on the Faroe Islands, and so was the white-speckled form, although never as numerous as the black form. But round about this time, the population started to shrink quite severely, partly because the raven was considered to be a vermin, and everyone had to kill a minimum number of ravens every year. The speckled ravens were also killed in large numbers, but in their case it was mainly because collectors and museums were prepared to pay serious cash for specimens for their collections. The last known speckled raven was shot in 1902, and today only 26 specimens can be found in various museums – 6 of which can be seen in Copenhagen, all of them with various amounts of white feathers around the head, wings and underside, and typically with a much lighter than normal beak. They are now in a rather sorry state, as a large number of researchers and writers have been allowed to see and handle them.

Since the killing of the possible last pied raven in 1902, there has been several alleged later sightings, in 1916, 1947, 1948, 1965, 1976, 1978, 1988, 1989 and possible in 2006 and 2008, but none of these birds were shot or photographed, so it is impossible to ascertain how reliable the sightings are. Some of these birds have been described as completely white, and this could mean there is yet another strangely coloured raven to be found in the Faroes, but I will return to that a little bit later.

Mutations
The exact nature of the pied ravens has been a matter of some debate. Most researchers think that is was some form of recessive mutation affection the production of melanine in the feathers. Recessive mutations can be quite difficult to study, as the mutated gene will only be expressed if the animal has two copies of the gene, where as animals with only one copy of the mutated gene will look normal. In the case of the raven this would f. inst. mean, that two normal looking ravens who were carriers of the gene, could have both black and pied youngs, a pair of pied ravens would only

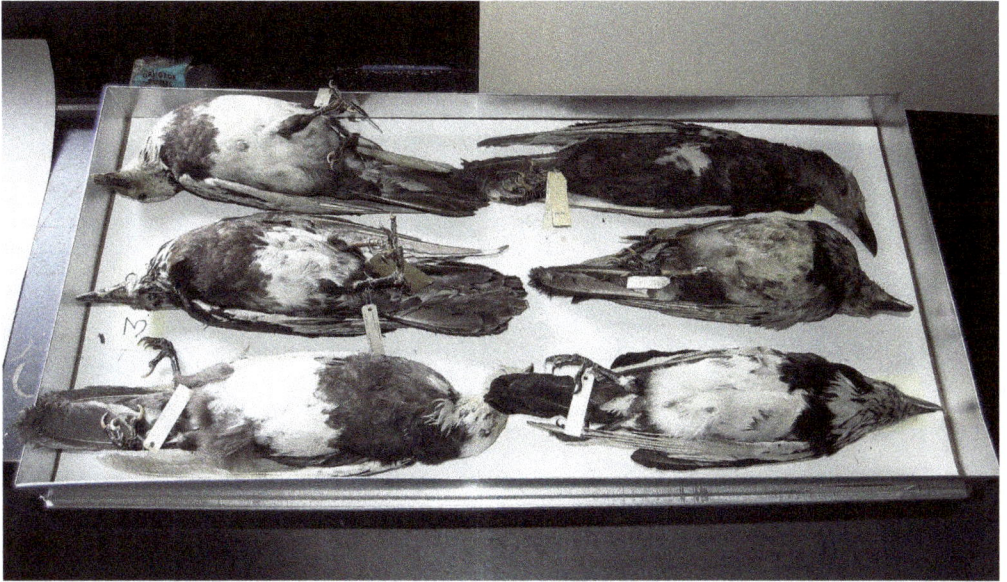

have pied youngs, but any pairs were at least one of the birds had two normal genes, would only get normal looking youngs.

It is of course a theoretical possibility, that some of the ravens in the Faroe Islands today are still carrying copies of the mutated gene, but it is highly unlikely. Since the population hit an all time low in the first part of the 1900's, it has grown steadily, and not a single pied bird has been seen. The mutated gene was probably exterminated at the same time as the entire population almost disappeared.

But for some reason the Faroes Islands are conductive to mutations. In recent years, another form of strangely coloured ravens has started to appear. They are not pied or speckled, but has a uniformly very light silvery grey colour. Several of these birds have been seen since the late 1980's, some have been photographed, and one bird has been shot and given to the Faroe Museum in Thorshavn, where it is on display today.

Strange things are definitely afoot in the Faroe Islands – who knows what will turn up in the future?

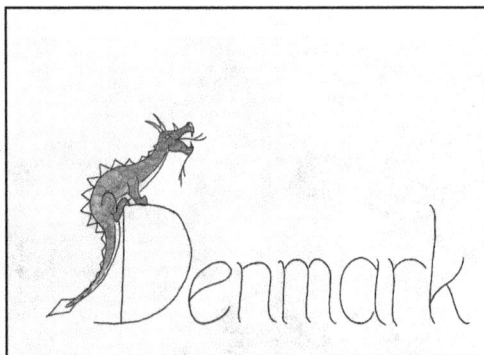

DENMARK - Small country – big stories

part from a few jokes about the Little Mermaid, the stories of Hans Christian Andersen, some bloodthirsty Viking stories, and the age of the Danish monarchy, the first thing that usually springs to mind when people talk about Denmark is the word *small* – and, let's face it, Denmark *is* a small country – even the Danes will acknowledge as much. It is, in fact, *so* small, that I have been asked by American tourists whether it is in fact a part of Sweden. It is also a rather— shall we say - disjointed country.

According to author Bill Bryson, Denmark looks like a plate someone has dropped from a considerable height onto a hard kitchen floor. The country consists of one fairly large peninsula and about 400 islands of various sizes. Denmark is also intensively cultivated and with a fairly dense population, making it difficult, but by no means impossible, for large animals to hide themselves. There are only a couple of large forests, and no mountains, but as the case of the Danish wolves has shown us, big things *can* hide here – and have done so on several occasions. In actual fact, there is a wealth of unexpected and surprisingly varied animals running around in Denmark, giving bored reporters something to write about in the middle of summer when the Danish parliament is not in session. In a sense, there would be absolutely nothing strange about most of these animals, were they only roaming around somewhere else. There is nothing wrong with a bobcat in a North American forest, but when it's seen repeatedly along a short stretch of bathing beach just south of central Copenhagen, it is far from ordinary. Denmark also has a rich and varied folklore describing all kinds of weird things – from small and insubstantial fairy-like things to trolls big enough to chuck enormous boulders around.

Those pesky wolves
The word wolf is one of those names which a strange and almost supernatural ability to install fear in the bravest of men. To hear the sound of a howling wolf pack in

Medieval Times while you were out and about chopping firewood or gathering mushrooms in autumn, must have been a truly terrifying experience. The wolf has always been an animal treated with a good deal of respect and fear, so much so that even the Vikings named some of their children Ulf (wolf) to bring them luck, and hopefully to see them grow up to be strong and brave, but also cunning warriors. But, as the last few years have shown in Denmark, this fear of wolves is still very much present.

Despite heaps of studies and the efforts of numerous scientists, which - among other things have shown - that you are far more likely to be killed by, say, a horse or a cow, far too many people still see the wolf through the eyes of Little Red Riding Hood, or have their knowledge of the wolf from various horror stories and movies depicting the wolf as a bloodthirsty monster that is best killed as quickly as possible. So you can imagine the panic, debate and sheer hysteria, when a birdwatcher in early 2012, almost exactly 200 years after the last officially recognized wolf in Denmark was shot, took a picture of an animal that looked very much like a wolf in a very isolated nature reserve in the North-western part of the country.

The end of the wolf

From a historical point of view, 1813 was nothing special in Denmark, but from a zoological point of view, it was in fact very much noteworthy, although for a sad reason. It was in this very year, on a cool night in June, that the last wolf in Denmark was shot, when it attacked a foal close to Estvadgård near Skive in North-western Jutland.

The story of the wolf in Denmark is long and complicated one. It arrived in the country about 13.000 years ago, and lived perhaps - not peacefully, but at least in a reasonable balance with the human population - until about 750 years ago, when the Danish king, Valdemar, decided that the wolf had to be eradicated from his kingdom. This set the stage for centuries of dedicated persecution of the wolf. It gradually became so important, that in medieval times, everybody had to pay a special wolf tax, to help pay the expenses of the very large specially arranged wolf hunts. And large they were. During a hunt in the southern part of the country in 1695, almost 4000 beaters formed a 50 km long beating chain. By 1513 the wolf was gone from the eastern part of Denmark, but it hung on for another 300 years more in the western part.

Unofficially though, there has been at least 120 sightings of wolves in Denmark since then. I have personally collected about 100 sightings, stretching from 1898 to the end of the 20th century, so I have no doubt that Denmark has been visited by wolves on a more or less regular basis. The animals have most likely come from the population in Germany and Poland. It is a well known fact, that young wolves especially, can wander enormous distances when searching for a place to settle down. Most of the sightings are from an area in the central part of Jutland around the town of Silkeborg.

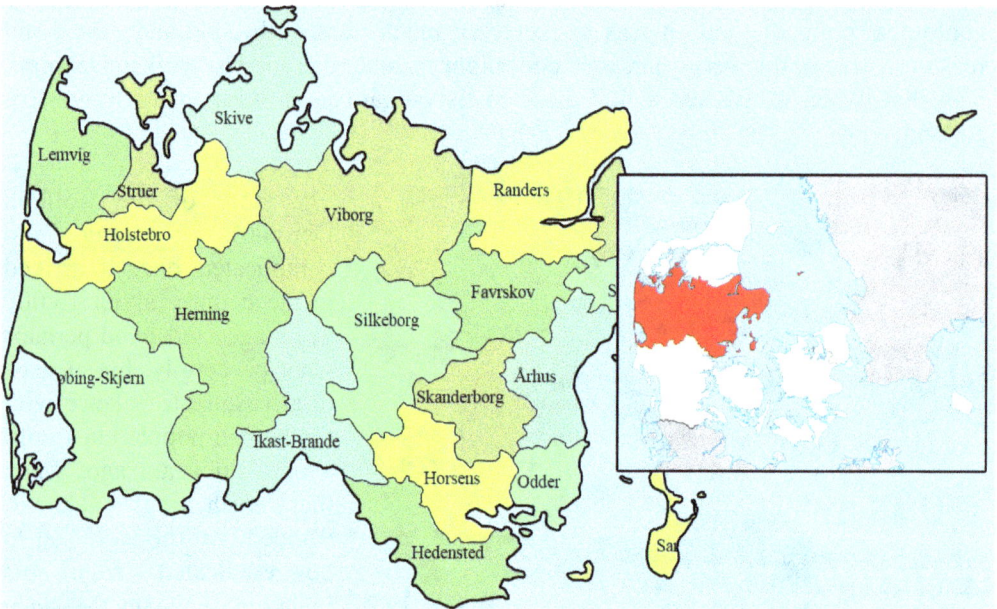

This is an area with a lot of lakes and streams, and lots of small and medium sized forests. It is also very touristy, but that apparently haven't deterred the wolves.

Just to give an idea of the types of sightings, here are a few examples from the last 30-40 years:

In the autumn of 1973, two people looking for a little fungal delight for their supper, saw no less than five wolves running across a pathway about 100 metres in front of them, as they were searching the forest litter in Store Hjøllund Plantation for edible fare.

On a winter's day in 1987, a man driving to his home in Silkeborg, saw a wolf running across open ground just south of Himmelbjerget, one of the most popular tourist attractions in the area. Round about the same time, at least five different people in the same area claimed they had heard a wolf howling.

In 1996 a wolf was seen near Stubbe Lake close to the Kattegat coast, and ironically, very close to one of the most intensely studied natural areas in Denmark, the socalled Mols Bjerge (Mountains of Mols), one of the first areas to be designated a national park.

In 1998 a wolf was seen close to the village of Tarp near the west coast of Denmark. .

An early morning in June 1999, at German family on holiday in Denmark, saw what they described as a very large and powerful grey dog with a long thick tail, standing at the edge of a forest not far from Salten Skov, which ironically is the home of the Salten Skov Laboratory, a research facility used by several Danish universities, and in summer usually filled with scientists and biology students. I have tried very hard to ascertain whether any biologist have heard or seen wolves in the area, but none of those I have interviewed would admit to a sighting. I spent most of a summer there in 1981, when I was a biology student at the University of Copenhagen, but did not see or hear anything unusual at the time.

In 2007 a wolf was run over by a car in Northern Germany, only about 100 km south of the border to Denmark, and in 2010 a wolf was caught on a wildlife camera in the same area, perhaps the overture to the later invasion.

In 2009 and 2010, there were several sightings in the areas around the border between Denmark and Germany.

In 2011 there were several sightings of wolves in the coastal plantation along the west and northwest coast of Denmark.

And then...

A new beginning?
Until 2012 there was not a scrap of tangible evidence for the actual existence of wolves in Denmark. I have a few descriptions of very large pawprints from the area, but unfortunately no photographs, and in 2003 a sheep was found killed in a way consistent with a wolf kill. Unfortunately I only heard about this dead sheep two weeks after it had been found, so it was not possible to find any traces of the killer.

The 2012 photograph taken in a protected area in northern Denmark, made quite a stir in the country. A stir that became a fully formed maelstrom when the animal was found dead (due to a cancerous growth in its throat, which had made it impossible for the poor animal to eat, so it had died from starvation) some weeks later, and proved without a shadow of a doubt, that officially for the first time in 200 years, Denmark was once again home to at least one wolf.

And of course, within minutes, angry farmers, scaremongers, conspiracy theorists and assorted idiots started coming out of the woodwork in incredible numbers.

A group of farmers and local politicians in western Denmark started an organisation dedicated to getting the wolf removed from Denmark before it started killing women and children and ruining farmers - in some of their press releases you got the distinct impression, that the last item was the most important. Of course people are entitled to their opinion, but I do prefer these opinions to be based on actual facts. Unfortunately said organisation very soon started distributing urban legends, hearsay and just about any horror story they could lay their hand on in an effort to scare the local populace into clamouring for the killing of the wolf/wolves. And I do say wolves, because in 2012 and 2013, loads of sightings suddenly started pouring in. Some of them were no doubt fuelled by hysteria, such as the wolf allegedly seen in a garden in the centre of a large town. But most of them were probably true, and indicative of the fact that Denmark probably was home to more than one wolf. And then an automatic camera caught an image of more wolves, and panic started to brew.

Detailed DNA-analysis of the first dead wolf, and later found droppings and other physical traces found around a dead sheep, showed that the animals came from a well-known population found in Eastern Germany and Western Poland. When this came out, the anti-wolf organisation started playing the conspiracy card. They knew, or had heard from sources they trusted, but of course couldn't name, that the wolves (first one animal, then two, and in the latest version of the story, three pairs), had been released deliberately (probably by leftwing environmentalist hell-bent on destroying the Danish agricultural industry or people just doing it for the excitement – although they had no idea as to from where they would procure six wolves), since it - in their opinion - was impossible, that the wolves had walked or run all the way from Poland to Northern Denmark without having been seen on the way. The only trouble was, of course, that they had been seen.

And then someone started a pro-wolf organisation and website, and everything went nasty. Hunters started to say that they would shoot any wolf on sight should they see it (of course just to protect their animals and children) and stuff the law. Oh, and - of course - the death threats started to fly. I personally received several, because I was all for receiving the wolves with open arms, whereas some people thought I was endangering their children and their animals by my reckless ideas. I offered several of them to take a walk anywhere they would name, but they never took me up on the offer.

And the number of wolves kept on growing. There is some dispute as to how many wolves there actually are in Denmark at the moment, but there are probably at least twelve different animals. One pair already bred in 2017, and will have probably done the same in 2018. But unfortunately, four wolves have disappeared without a trace, and one has been shot by a local, right in front of several people with cameras.

It looks like the wolf has come to Denmark to stay, but—apparently—so has wolf hysteria.

The weird ones

Should one have a couple of spare hours on hand, and a pile of sightings of one particular animal/animals. One has a – I think the technical term is a flap; it is a good learning experience to go through them rather thoroughly, because in most cases one will find a small handful of very strange ones, that usually crop up when the flap is well under way, and people are seeing whatever creature we are talking about left, right and centre, Such was indeed the cases during the wolf flap in Denmark in 2012 and 2013. It started out with a few scattered sightings here and there, and then suddenly the avalanche – a host of sightings within a very short time, when people who had already seen the wolf/wolves, but had been afraid to tell their stories, suddenly dared to speak up. This of course started the panic part, where other people saw wolves everywhere and started to mistake other animals and everyday objects for wolves. At first Alsatians and similar dogs were playing the part of wolves, and then smaller and smaller breeds, as well as cats started to come into the equation, as well as a rolled up carpet and a pile of shredded tires.

After a couple of weeks of this, the weirdness button was turned up a notch. Now the wolves started acting strangely. People claimed to have seen them in their own back gardens, running across streets late at night or early in the morning, and in one case a "wolf" was even seen on a walkway outside a row of 6th. floor flats in one of Denmark's larger cities. Oh yes, and one lady claimed to be in telepathic contact with the wolves, and had seen what she called the shadow of the wolf-mind in her living room every time a message was coming through.

Towards the end of 2013, things started to die down – the number of sightings dropped a bit, the most radical scare mongers disappeared as suddenly as they had appeared a year earlier, the anti-wolf organisation closed their website, and everybody now seems to have accepted, that the wolf is once again part of the Danish fauna.

Or perhaps not…

In December 2012, a dead fallow deer was found in a deerpark just north of Copenhagen. It was rather obvious, that the animal had been killed by a large predator. It has a nasty bite on its behind, and its head and neck was missing. Within seconds, everybody was screaming "Wolf!" and wanting every wolf in Denmark shot at once, because now it was only a matter of time before a child would be killed. Oh yes, and the wolf had been deliberately released – as the deer park is completely fenced in – or the deer was brought in from somewhere else and deliberately put in the deer park to judge people's reaction as a preparation to the actual releasing of wolves.

Luckily the local forest warden kept his wits about him, and arranged for DNA-samples to be taken from the dead deer to determine the actual identity of the killer. It was a dog!

A cornucopia of animals far from home

As any birdwatcher will know, you sometimes come across birds very far from home; birds whose navigational systems must have had a major crisis/breakdown, since they have ended up thousands of kilometres away from where they should be, often having flown in the opposite direction of where they were supposed to have gone. Most of these sightings can be explained, and the same goes for whales that get lost somehow and other animals with wanderlust. And then of course there are those where your only reaction is something along the lines of "What?"

Such as the Australian emu, that was found drowned in the 1970's in Western Denmark. Nobody came forward to claim the bird. Or the similarly life-challenged cassowary found dead in roughly the same area about 10 years later. There is also a good selection of kangaroos and wallabies, as well as racoons, llamas and the odd leatherback turtle. And then there are the really interesting ones...

My sister was bitten by a moose...

The moose is very much a thing of the past in Denmark – and then again, not really. Although the biggest deer in Europe once was a part of the Danish fauna, it no longer

is. But on the other hand, there are plenty of moose in Sweden, and likewise in places like Poland, so why not in the future – why not indeed? In actual fact, a small group of mosse has already been released in an enormous enclosure in Northern Denmark – and has already produced a couple of calves. And in Almindingen Forest on the easternmost island of Bornholm, a herd of European bison has been disporting itself for several years, and has already produced several calves.

Meanwhile, the moose – especially the Swedish ones, have apparently decided, to do something about the fact, that they no longer live in Denmark. I don't know if they draw lots, or that every now and then, a Swedish moose suddenly decides it has had enough of cheap pop music, and decides to leave.

And here comes the interesting thing – moose are excellent swimmers, and I do mean *excellent*. In spite of their rather strange bodily shape, and clumsy and ungainly way of moving, they can swim for miles, and even dive. And they demonstrate it at a fairly regular basis, by turning up – usually in North-eastern Denmark – much to the interest of local people. And this entails a swim of at least 4 kilometres through an area with very strong currents. In most cases the animals have probably had to swim at least twice that distance. Some moose never make it – in 1997 a dead one was found at the north coast of the easternmost Danish island of Bornholm. But usually the moose settle down fairly happily. One especially friendly one lived for years in the 1950's in Grib Skov, one of the largest forest in Denmark about 25 kilometres north of Copenhagen making friends with the locals in the villages in and around the forest, and in some cases helping itself to the fruit and vegetables thoughtfully laid out in front of the greengrocers shop in the village of Nøddebo in the middle of the forest.

Most of the moose have come to an unhappy end though – especially in the last decades of the 20th century – where they have all ended up being killed by cars or trains.

Turtlepower

A lot of countries in Europe are home to free-living turtles of various kinds. Most of these are descendants of North-American terrapins or similar species that are common in the pet trade. Many people buy them, but when they get tired of them, it is apparently easy just to drop them in a lake or a pond somewhere, and forget all about them. In some areas this is incredibly common.

The Botanical Gardens in central Copenhagen has among its many features a small lake, and so many people use this lake for depositing unwanted animals, that the gardeners has to perform catch-and-remove operations a couple of times a year to keep the population down. And it is not just the more common species – much rarer Chinese (pictured above) and other Asian aquatic turtles have been sighted and photographed in just that small lake – by the last count six different species all together. Just to give an example – in the summer of 2013 one could see a Chinese three-keeled turtle sunning itself on a rock close to the bank of said lake.

So when people in the 1990's, started talking about European pond turtles in various lakes around the town of Silkeborg in Western Denmark, it didn't exactly make

sensational headlines. Officially the pond turtle had died out in Denmark round about the time of the bronze-age, so the general consensus was that the animals found were pets that had been released. But were they now? First of all, the pond turtle is not especially common as a pet. It is considerably less colourful than the American red- and yellow-eared terrapins perhaps, and so not as interesting to keep. But a bit of historical research showed that one could find sightings of the turtles in the same area going back for at least a 100 years, way before the keeping of exotic pets became popular.

So, although most people thought the turtles were nothing more than escapees, in 1996 it was decided to perform a DNA-analysis on five animals that were captured alive in the area. The results were compared to animals from various known pond turtle populations in Europe – the nearest being in Poland. It turned out, that four of the Danish animals were genetically identical to the Polish animals, which could only mean, that they were escapee pets or descendants of escaped pets, as there was no way wild-living Polish animals would be able to make their way to Denmark, and it was a known fact, that at least some of the animals you could buy in the pet trade, were either caught or bred in Poland. But the last of the turtles had quite a different DNA-profile. It did show some similarity to the Polish animals, but it also differed in several ways. Some researchers suggested that the differences could be explained by

the fact, that the animal was part of a population that had lived on, hidden and isolated, since the end of the Bronze Age, and had gradually started to separate genetically from its ancestors.

No further DNA-research has been done on the turtles, so nobody quite knows what to make of them. But subsequent research has shown, that there is in fact a small, but apparently viable population in the area around Silkeborg, but only time and further research will tell, whether they are in fact all just feral turtles, or a real 2000 year old relict population.

Here Kitty...
Alien Big Cats, or ABC's or simply Big Cats, are undoubtedly one of the best known type of cryptozoological animals. There have been sightings of big cats just about everywhere you can think of – and here I am of course talking about the countries where big cats are not a natural part of the fauna. Denmark is no exception. There have been hundreds of sightings of several different species. They could easily fill a book all by themselves, but that would be boring in the long run, so I have made a selection of interesting and representative cases.

By far the most common big cats reported in Denmark are the puma, and the lynx, but we also have sightings of lions, although they were probably golden cats slightly exaggerated in size, servals – in this case a real animal, a former pet that escaped and managed to live and roam freely for several months before it was recaptured and reunited with its family. Strangely enough we are rather lacking in the black cat department – I have only managed to find two of those. One was a straightforward black panther that escaped from a zoo in Northern Denmark, and was finally cornered in a block of flats in the town of Aalborg. The other one was/is considerably more strange and scary... and spooky!

The Gorge Cat
As I have mentioned before, cliffs, mountains, gorges and similar wild landscapes are in rather short supply in Denmark – or if we have them, they are of decidedly modest dimensions. But on the north coast of the island of Bornholm – the easternmost part of Denmark, you can find something that at least in Danish terms, is a dramatic gorge. It's called Randkløveskår, and you find it a little bit west of the little town of Svaneke. It is a 12 metre deep and several metres wide crack in the rocks. It looks like some ill-tempered giant has been laying about him with a large axe. It is a strange place, and strange stories have been told about it.

It is one thing getting a little jumpy when you start to crawl down into the bottom of the gorge, and perhaps start to feel a little claustrophobic as the walls begin to close

in. It is considerably more difficult to explain what has happened, when people claim to have seen the luminous eyes of the Gorge Cat down at the bottom. I have been clambering around in that very area on several occasions, and have unfortunately never seen anything strange – maybe I don't look appetizing enough, or maybe I am wheezing too much. Although the gorge is of a modest size, it is quite a hard climb to the bottom, and especially up again.

During Easter 1983 L.H. was on a bicycle trip around Bornholm with two of her friends. One day they had planned on driving from Svaneke to Gudhjem, the next little town along. None of them knew what Randkløveskår was, but when they passed the turnoff, and saw the sign, they liked the unusual name, and decided to investigate. Having admired the views from the top of the gorge, her friends lost interest, and decided to find a suitable place for a spot of lunch. L.H. on the other hand was deeply fascinated by the place, and started descending to the bottom.

It was a rather grey and cloudy day, so when L.H. reached the bottom, it was considerably darker than she had anticipated. She had just turned around to look up on the grey sky and try to call to her friends, when she heard a scratching noise behind her, and got the distinct feeling of not being alone anymore.

"I was seized by an almost uncontrollable desire to scramble up the rocks in a blind panic, but for some reason I managed to stay put. I kept repeating: Relax, you're dreaming, relax, you're dreaming! And the whole time I could hear a gentle wheezing sound, as if something or someone was breathing quietly behind me.I took me ages to gather enough courage to turn around, but to this day [I interviewed L.H. almost ten years later], I still wish I hadn't. It would have been easier, if it had been some kind of local twat standing behind me, but it wasn't. It felt like darkness had suddenly filled everything around me. I had the feeling I was looking through a door or a window into a completely dark room. Within a few seconds I realized that the

"room" was far from empty. Now I could clearly see two glowing green eyes in the darkness. It was the eyes of an animal – a large animal! I was completely petrified. I don't know how long I stood there, but gradually I realized I was looking at a very large black cat. That broke the spell. I turned and ran screaming up the rocks, skinned my left knee, tore my trousers and sprained my wrist. It hurt, but all I could think of was putting as much distance between the cat and myself. When I reached the top, I just ran past my confused friends, jumped on my bike and started pedalling like crazy towards Gudhjem. I only calmed down 3 or 4 kilometres down the road, when I ran out of air. I have never dared to go back, and I still have nightmares about it sometimes."

A stray cat or an overactive imagination? Who can tell? According to some locals, the bottom of the crack doesn't always end in bare rock, but leads to the entrance of a dark tunnel leading way down into the bowels of the earth. As far as I have been able to ascertain, nobody has ever dared to enter the tunnel when it has been open, or at least haven't lived to tell the tale. So whether it is all an illusion, or whether it is an entrance to another world, guarded by the Gorge Cat, is anybody's guess.

The cats with no tails
In the early 1970's, when all was well, and the world had money aplenty, the Danish government started building a series of marinas, artificial lakes and beaches a few kilometres southwest of the centre of Copenhagen. It was a project of quite colossal proportions, and apart from the sailing and bathing possibilities, a lot of people had high hopes for the area in terms of natural history. The birds arrived almost instantaneously, plants started turning up and so did insects and one rather surprising mammal.

The first stories about strange animals started surfacing around 1975, when the area was still under construction. At this stage it was a strange and otherworldly landscape. I was fascinated, and as I lived quite close by. I visited the area on a regular basis. You had to move carefully, as the constant digging and repositioning of tons of sand, created intermittent pools of quicksand that changed position almost daily.

The first people to realize that strange things were afoot, or possibly apaw, were joggers who sometimes, especially early in the morning, saw a big grey animal running around the sanddunes, or dogwalkers whose dogs, sometimes even big and strong ones, started whimpering and hiding behind their owners, absolutely refusing to pass certain points. And then of course there were birdwatchers like me, who insisted that they had on several occasions seen what can only be described as a

medium-sized cat, 1-1½ metres in length, grey and brown with indistinct spots, large powerful back legs, large black tufts of hair at the tip of their ears, and a short and stumpy tail. You don't have to look for long in various books of animals to find a species that suits this description to a t – a lynx. Unfortunately nobody was ever able to get a photograph – the animal was usually only seen in fleeting glimpses – but man did we try. At least I know I did. It cost me and arm and a leg, but nothing ever came off it. And nothing ever came of the lynx. I have no idea where the animal came from – it must have been somebody's pet. Although lynxes have a wide distribution in Europe, a wild one would have had to manage one hell of a swim to get to where it was seen.

The lynx disappeared just as suddenly as it had appeared. As the beaches and marinas got nearer and nearer to completion, more and more people started coming to the area, and I presume it just got too much for the lynx at some stage. It probably tried to find somewhere with more peace and quiet. Perhaps it made it, and lived out its life in true lynx-style, or perhaps it was hit by a car, limped away, and died in some forgotten corner somewhere. I really have no idea. But what I do know is that it was not the first lynx in Denmark, and it definitely wasn't the last one either. There are in facts heaps of sightings, but I will only describe one other case, as this was quite a big thing.

The deer-killer

In 1996 someone new to the district moved into the peaceful Stenderup Forest a few miles north of the city of Kolding in western Denmark. Stenderup Forest is not a big place. It is actually three different small forests and a few bits of open ground, comprising about 8½ square kilometres all together. It had a substantial population of roe deer, but from 1996, they started taking a serious beating, when a predator big enough to bring down a fully grown roe deer, started killing them with what can only be described as gay abandon. Forest rangers, dogwalkers, and people out for a stroll started finding roe deer with broken necks on a regular basis.

Media-interest was slow at first, but started getting momentum in 1997, and reached a slightly breathless and mildly hysterical zenith in 1998, when so many people started visiting the area that the chief ranger had to ask the newspapers to pipe down, and people to go home. Several attempts were made to catch or kill the animal was made, but to no avail, and gradually sightings of the beast, undoubtedly a lynx, as well as the number of killed deer started to drop again, until one day everybody realised, it had either died or moved on. Unfortunately DNA-analysis had not come into general use at the time, and I was never able to convince the forest people to try and get some hair-samples, so we never got anywhere near a definite id for the perpetrator.

The story doesn't quite end here. In the following months t here were a couple of sightings further north, of what I think was the same animal, but that's it – end of story, death of lynx I presume.

The green kitten

The last cat in this section is also a bit of a mystery, but on a whole different level. There is absolutely no doubt about its existence. It was photographed and studied and featured in all the major newspapers in Denmark. It was to all extent and purposes an ordinary housecat – or actually a kitten – apart from one minor detail. It was green!

The kitten was found on a farm in northern Jutland in the autumn of 1998, and it made quite a stir in Denmark. Dubbed 'Miss Greeny', it did enjoy a brief period as a national celebrity. Its siblings were a standard grey colour, but Miss Greeny had the same colour as an old copper roof. Samples of her hair were sent to the Danish State Hospital in Copenhagen, hoping they could discover the cause of the unusual colour.

There were rumours of hoax, of spraypaint and hairdyes, and God knows what else. But the researchers did not find any trace of artificial colouring. The kitten's hairs were green, from the tip to the roots, and the colour was in the form of green pigment inside the hairs.

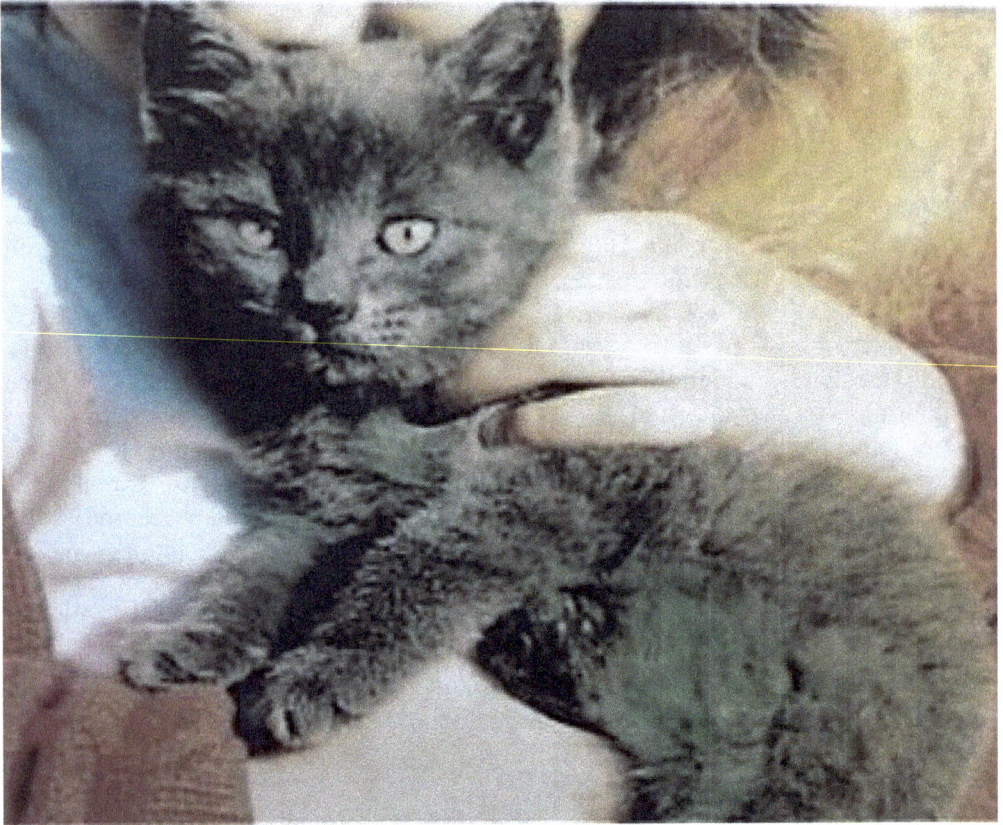

Two theories were put forward at the time, none of which could be proved or disproved. One was that the kitten had drunk water with an unusually high copper-content, and this had somehow found its way into the pigment producing processes in the kitten – which altogether seems unlikely, as none of Miss Greenys sisters or any of the other cats on the farm were green – not even a little bit.

The other theory suggested that Miss Greenly had some form of genetic disorder, resulting in the addition of copper to the normal pigment produced in the hairs.

In the end, the whole thing kind of petered out, as the kitten lost the green colour, as she grew up, and spent the rest of her life (14 years) as a normal coloured grey cat. Naturally, the family on the farm have kept their eyes out for any more green kittens, but so far it appears that Miss Greeny was a one off.

And so, Miss Greeny disappeared; quietly and stealthily as is the custom of cats. In fact sometimes they come and go in a way that makes you wonder whether they actually have a solid physical form. They do, of course, but whether the same goes for the next group of creatures is a decidedly different matter. There may be genuine flesh and blood animals behind some of these stories, or we may be dealing with some form of ghosts, or perhaps what has been termed zooform phenomena, animal-like manifestations of something – perhaps some kind of primeval energy... or perhaps not. Anyway, time for at look at:

Fiery dogs and flaming horses from Hell

A large proportion of the animals I have described above have a clear physical reality. They may be misidentified, they may be the result of pranks, they may be escapees or they may be something else entirely. In Denmark we also have a fairly large proportion of beings whose existence is a bit more questionable or they may not be altogether physical - at least not all the time. We are talking about animals that one moment can seem perfectly solid and normal, and then disappear or dissolve in seconds, or perhaps walk through walls, levitate or sport eyes and mouths filled with flames or glowing red or some other ghastly – or perhaps that should be ghostly – supernatural colour. The most common example of these beings are the black dogs – for some reason especially common in Great Britain – but also fairly well-known in Denmark. We have a good selection, and they seem to be a bit more energetic than their British counterparts, as the Danish black dogs tends to patrol rather large rural districts, compared to the often measly lanes sought out by the British spectral hounds.

Apart from the dogs, Denmark is also home to a host of other semi-spectral beings serving various dark and nefarious purposes. So if you are out and about late at night, you may run into things like grave-sows, corpse-lambs and hell-horses – collectively known as church-warders.

Never close the door on a black dog

The black dogs of Denmark are quite active compared to black dogs from other parts of the world. Those usually just haunt a stretch of road or a particular location, but the Danish versions are rather more busy. Some of them have to patrol enormous areas every night, running long stretches along certain roads, often with blinding speed, in and out of farmyards, gardens of manor houses and royal parks. And because the dogs are in such a hurry, God help anyone who close their gates at night, thus obstructing the dogs progress, and when that happens, it can only end in tears. The dogs do not take kindly to these kinds of things, so you can count yourself lucky if you get off with a gate blown to splinters. In worst cases the dog will manhandle you so severely that you can never walk straight again. But strangely enough, if you couldn't be bothered to open your gates every night, it was sufficient to drill a small hole in the gate. This is perfectly sufficient for the black dogs, although some of them have been

described as being the size of a small pony. This tradition persists to this day in several areas of Denmark, with small holes being made in gates and doors.

But of course the dogs doesn't just run. They have duties to perform as well. They have to show themselves to people who are about to die – (usually people who have behaved rather badly in life) – and they have to pick up their souls if need be, and take them straight to the warmer parts of the afterlife.

The patrolling dogs are an efficient lot. When they have to pick somebody up, they don't bother with preliminaries, they just barge in, often to the tune of exploding windows, pick up their intended target, and leave just as suddenly and violently, leaving nothing behind but a faint whiff of sulphur.

Pejter
Some of the dogs are a little bit more laid back and relaxed, and also demonstrates a much more friendly side. On the archipelago Hirsholmene in northern Denmark, there is a black dog locally named as Pejter (local dialect for Peter). It is a very large black and shaggy black dog, and it looks rather terrifying, seeing it has red luminous eyes and similar coloured maw – in some cases even spitting fire. But no matter, Pejter is

actually rather friendly. He does in some cases show up to gather the souls of the departed, but mostly he guards the islands and makes sure nobody gets harmed before their time has come. There are plenty of stories of people meeting Pejter late at night, but not feeling the least bit scared – in some cases Pejter even leads people safely home. On one occasion in the winter of 1942, he even intercepted two people who had blundered on to the ice surrounding the main island of Hirsholm, and had made sure they found their way back to dry land, and not tried walking to Sweden.

Hirsholmene was never a big place – in the 1800's, there were some 225 inhabitants, and this is also the time from where we had the most sightings of Pejter. Today there is only a small handful of permanent residents on the main island, but Pejter is apparently still around. In the summer of 1981, two students from Copenhagen claimed to have seen a big dog with glowing eyes disappear behind the island's tiny church. But they couldn't find any trace of it, when they started looking for it. And I must confess, I had a hard time believing them, when they told me. Both of them, and me, were part of the same group of students that visited Hirsholmene as part of a stay at the Marinbiological Laboratory in Frederikshavn in Northern Jutland. There was not a word spoken about dogs when we were there. The only told me several years later, when I was working on my first book, and told them about my interest in those kinds of stories.

The Moesgaard dog

If you go to Århus, the second largest city in Denmark, you may run into a black dog of a considerably nastier disposition. In the outskirts of Aarhus, you can visit, if you so wish, the Moesgaard Museum. It is rather splendidly situated in a very beautiful park, and it is in fact in the park you can run into a big black dog, but this is not an experience you should long for, as a meeting with the Moesgaard black dog brings you nothing but sorrow and despair. If you are not dead or severely ill, or stricken by the most devastating bad luck within a few days of a meeting, someone close to you surely will be. Most of the sightings of the Moesgaard black dog are from before the First World War, but a few are actually of a later date.

In 1992 a girl who worked at the museum was walking home through the park early one evening, when she suddenly came face to face with a large black dog that just stared at her. She tried to take no notice of it, and just walked past it. About 25 metres further on, she looked back, but the dog had disappeared. She got home, told her boyfriend about her meeting, and went to bed. The next morning she was violently ill. A doctor who was summoned, was unable to diagnose her illness. Within a couple of days she was so ill, she had to be hospitalized. About a week later she started feeling a bit better, but it took almost two months before she was back to normal.

A wealth of wardens

Church-warders are creatures of a strange nomans- or perhaps nosouls-land somewhere between the living and the dead. They are the results of the last pagan beliefs spilling over into the start of the Christian era in Denmark. From the end of the Viking-period until somewhere well within the Medieval, churches were going up right, left and centre in Denmark. The churches were often built by fairly small local communities, usually encouraged by a monk or missionary from other parts of Europe.

All though the Danes embraced Christianity quite willingly, they were a bit reluctant to give up all their ancient beliefs in one go. You never know – perhaps some of the old gods would be angry. So to appease them a sacrifice was usually made before a new church went up, and sometimes other major constructions. It was usually an animal, and it was sacrificed in a rather gruesome way. The people would chop off one of the animals legs, and then bury the thing alive – or perhaps more like half dead – usually under the corner of the church or somewhere in the churchyard, all in the belief that the poor thing in that way would end up in a kind of limbo between the normal world and the afterlife, or perhaps existing in both worlds at once. For some reason lambs were especially popular for this. In this way the animal could keep the dead in order, make sure the souls stayed put until Judgment Day, and go about gathering its next costumers, and generally make sure everything was in working order in the churchyard.

Usually church-warders would only show themselves to people who were about to die, but sometimes casual bystanders would get a glimpse of them as well, often mistaking them for ordinary animals, before they realized something was very, very wrong. These sightings have continued all the way up to the end of the 1900s, although with a decreasing frequency. Today most of the Danish church-warders have been worn down to nothing more than a few fleeting glimpses, or the faint sound of hoofs on cobblestones or trotters on gravel. Some black dogs are also confined to churchyards, where they "work" as wardens, as I have described above – but they have some rather interesting co-workers in that particular field apart from the lambs.

On the central Danish Island of Funen, grave-sows are for some reason particularly common – and most of them are red! In western Denmark, in Jutland, corpse-lambs are the standard church wardens. They can often be recognized by their extremely long an unkempt fleece. Several people have seen these strange lambs in churchyards – in one case as late as 2004. That one got reported to an animal welfare organisation, but when they arrived to take the lamb away, it had disappeared into thin air.

One of the church-warders can be found in very illustrious surroundings. The hell-

horse not only haunts the area around Roskilde Cathedral about 30 kilometres west of Copenhagen, where most of the modern kings and queens of Denmark are buried, it has its actual lair inside the cathedral. In the apse of the cathedral just behind the main altar, a row of gravestones have been sunk into the floor. The one on the extreme right is unmarked and completely black. Legend has it that below here is where the hell-horse rests in between missions.

Unfortunately the archaeologists have taken a peek under the gravestone, and found not the slightest trace of a horse skeleton under it.

Zooform phenomena
I have no idea was these ghostly beings are. They are clearly not physical animals, although some of the sightings attributed to the dogs and various other church-warders have in fact been ordinary animals seen under less than ideal circumstances. So are they some form of manifestations of latent spiritual or earth energies, or are they simply figments of people's imaginations? At this stage your guess is as good as mine.

Museum of monsters, or monsters in museums

Cryptozoology is not something one has to pursue in steamy jungles and windswept mountains. It can just as easily be done in old museums and curiosity cabinets as well. If they museum is old enough, there is almost guaranteed to be dragon bones or similar objects dating from the 16 – 1700's. Lots of things that we today know is straightforward elephant bones and various forms of horns and antlers were once classified as bones of giants or claws from various monstrous birds. Several of these things can be found in Danish museums. In Copenhagen one can, should one so wish, peruse a sizeable collections of unicorn horns (narwhal-tusks), griffin claws (chamois horns) and even some claws from a rukh-bird (ibex horns).

Denmark has traded with Arctic areas since back into the Viking-ages, so the narwhal tusks/unicorn horns have always been fairly common here, and fairly commonly used and studied. In Rosenborg Castle in

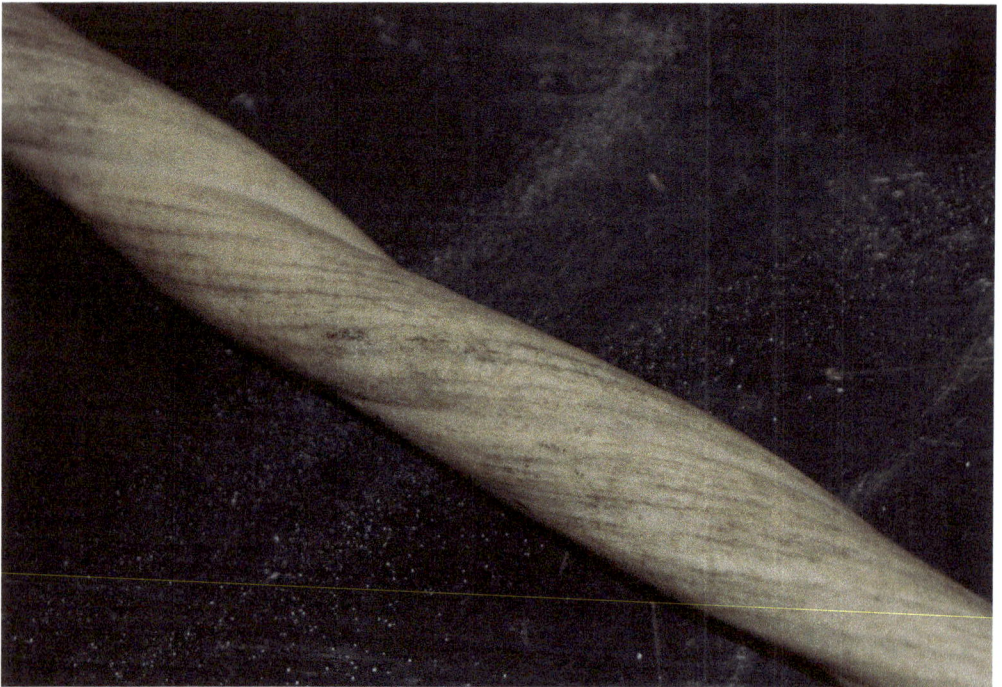

Copenhagen, there is a throne heavily ornamented with bits of unicorn horn – something which in its time must have been perceived as a symbol of immense power and wealth.

Several Danish scholars have written learned treatises on the unicorn and its horn, but also rather quickly realized what the so-called horns actually were. But they did nevertheless show true scientific spirit, and tried several experiments to wee whether the various miraculous powers ascribed to unicorn horn were indeed true. In 1636 an experiment was made in Copenhagen to test whether unicorn horn could in fact neutralize poison. A servant was given a small dose of poison ("not enough to kill him, should the experiment fail, only make him ill for some days. He is a perfectly good servant, and there is no need to let him go to waste like that") mixed with ground up unicorn-horn. The scholars, much to their own surprise, did in fact get a positive reaction. The servant only reported mild discomfort, although this may in fact have been caused by the fact, that the poison and the ground up horn had been dissolved in milk.

Scaly things
The natural reptile fauna of Denmark is very small. Two snakes – the grass snake and the adder – (there used to be two more, but they have been extinct for many years). Two lizards – sand lizard and common lizard, and the slowworm as well as the above mentioned European pond tortoise. We also have the odd sea-turtle showing up every now and then, but that's just normal albeit navigationally challenged individuals. That should be all, but of course it isn't. There are a number of sightings of other reptiles, going from large snakes up to and including fully grown dragons, although most of the Danish dragons are of a rather special local species called the "lindorm" – this is an old Nordic word meaning snake, and lindorms are very snakelike, only rarely do you hear about legs or wings, although some of them are thusly equipped and can fly. Some of them are also aquatic, but most prefer living on dry land, where they have a troublesome tendency to destroy churches. And we do in fact have a few full-blooded dragons with legs and wings and the whole fire-breathing scenario, although compared to the dragons of other countries, the Danish dragons are a rather sad lot.

Here be dragons
Dragons are big, majestic, usually ill-tempered and with a strong liking for gold and treasure – at least according to Lord of the Rings and similar forms of literature. In the Far East dragons are equally mighty and powerful, but also wise and mostly

benevolent.

In Denmark dragons are usually on the small side, timid and rather useless. And even so, we have stories about gallant knights fighting dragons – even St. George make an appearance here and there, although he is called St. Jørgen in Danish.

The Broager dragons

In the extreme southern part of Denmark, very close to the border with Germany, about 6 km west of the town of Sønderborg, one finds the considerably smaller town of Broager. It is not especially different from most of the other little towns in the area – that is apart from a rather magnificent white church with two imposing and very closely spaced spires. That is all connected to a rather gruesome story of a couple of conjoined twin sisters and their ghost(s), but we won't go into that here. What we will go into is a rather strange wooden statue located inside the church. It is a large wooden effigy of St. Jørgen (St. George) busy doing what he apparently did best – killing dragons. Not terribly heroic by the look of it, as the dragons is smaller than the knight's white horse. The statue is a reconstruction from 1880 of the original and much older statue. Actually, it is so old nobody knows its exact age, although we do know that Swedish soldiers in 1658 tried to drill a hole in the dragon and the horse looking for hidden treasure. There used to be a princess as well, by the way, but she is

long gone.

What really interests us here though, is the depiction of the dragon. It is very much a female, with two rows of well developed teats on its belly – which would suggest, that it was in fact a mammal, and as such rather unusual, as they are normally thought of as being reptilian. Unfortunately, there are no records of the origin of the statue, so we don't know what inspired this rather strange anatomy. But what we *do* know is that there have been two actual dragon sightings in the area around Broager, although both are about 200 years old.

The first story was told to me by a local historian. It concerns a local man who sometime in the 1820's had been out fishing from Iler Beach which is about 3 km southwest of Broager. Whilst rowing back towards the beach he chanced to look up, and saw what he said looked like a long glowing rod flying very quickly above his head from west to east. The thing – which he called a dragon – must have been very high up, because he couldn't hear a sound, but as he was watching, the dragon seemed to grow in size and glow even brighter, and suddenly started to spit fireballs in all directions before disappearing in a flash.

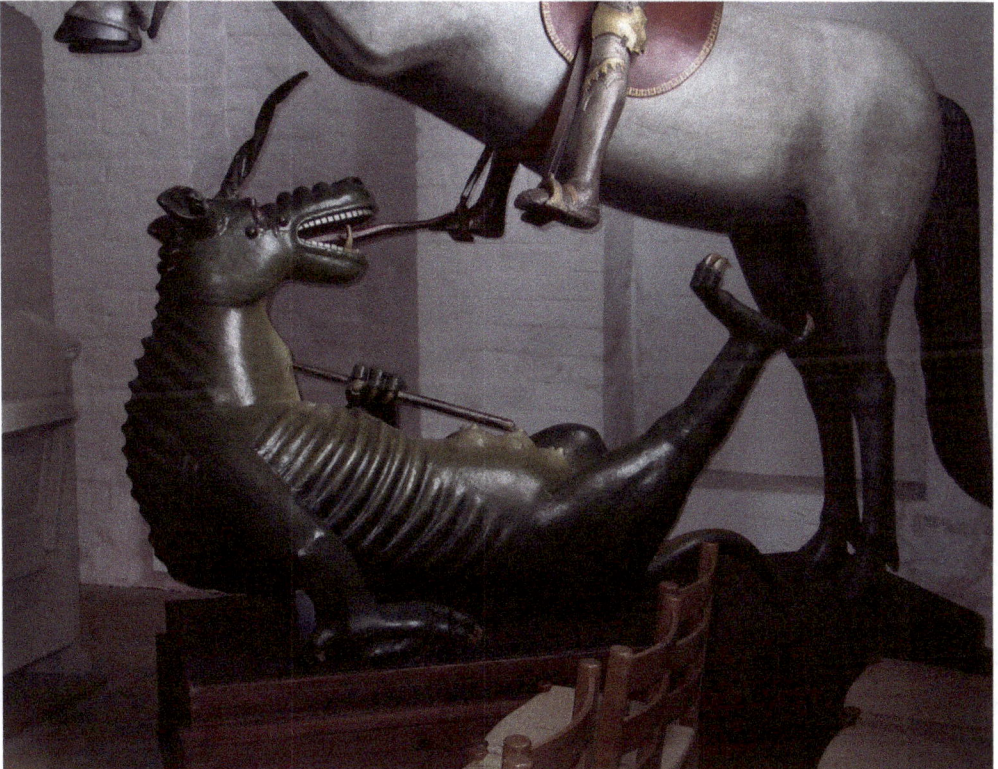

Now, I don't think you could find a better description of a meteorite breaking up in the atmosphere if you copied it from a textbook, so I think there is absolutely no doubt, that this was in fact exactly what this dragon was.

The other sighting is a bit more interesting and mysterious. It is a sort of family legend which has been told for several generations in the Hansen family. I came into contact with them in 1998 when I was in a school in a town called Rødekro giving a lecture. When I was chatting to the student afterwards, one of them told me that they had always told a story in their family about one of their ancestors having seen a dragon 200 years earlier. I got his parents address, wrote them a letter, and a couple of weeks later I received a written version of their family legend. It is rather long, so I will only give an extract of it here.

The family was very interested in genealogy, and had researched their family history quite extensively, and had managed to put a name to their dragonwatching ancestor. Apparently a distant uncle by the name of Hans Peter Hansen, had been living in the town of Sønderborg round about 1800. One summer's day, to the best of their knowledge in 1801, he was working in the Kobbelskov Forest about 7km southeast of the above mentioned Broager. Uncle Hans was a carpenter by trade, so presumably he was out looking for suitable materials to work with. Whilst chopping away, or whatever he was doing, he suddenly heard a strange rushing sound that grew louder and louder as the tops of the trees in the forest started whipping back and forth like grass in a hurricane. Looking around to see what has caused this disturbance, poor uncle Hans was scared out of his wits – according to the story he actually wet himself – when he suddenly saw what he described as an enormous lizard the size of a large cart with four horses in front, rising slowly from between the trees a couple of hundred metres away from him, and then starting to fly in a large lazy arc above the trees, before it banked and disappeared in a south-easterly directing across the sea towards Germany. He couldn't see any wings, so the dragon must have looked more like one of the oriental snake-like dragons, but whatever it's species and modes of propulsion, it was gone in seconds. Thus endeth the story, and we can basically just sit back in wonder.

The farting dragon
Somewhere above I wrote that Danish dragons by and large are a fairly useless lot, but none of them probably more so, than the farting dragon of Kalø. It lives deep below the ruins of Kalø Castle in eastern Jutland where it guards an enormous treasure, big enough to pay all the taxes in Denmark for a seven year period. All the money is to be found in a giant copper kettle, around which the dragon has curled itself. A lot of people have tried to dig down to the dragon's lair, but only twice have people managed to actually open the gates to the dungeons. The first one was an

adventurous German, but in his case the dragon acted suitably dragon-like and tore him to pieces. In the second case though, things did not start out well for the dragon, maybe because it was two local lads. In any case they spoke to the dragon in such a stern way that the poor creature got quite scared, and tried to hide in a corner of the cave. The two men were just about to start dragging the treasure op to the surface, when something rather unexpected took place. In pure fright the dragon released a mighty foul-smelling wind that brought tears to the eyes of the men, curled their hair and singed their shirt-tails. Coughing and spluttering they staggered towards the surface, leaving the treasure behind. And since then not a single soul have dared to enter the lair of the farting dragon.

The lindorm

The most common form of dragon in Denmark is the lindorm. It is shaped like a giant snake, but since the lindorm is mainly aquatic, I have written quite a lot about them in my book *Weird Waters* (CFZ Press 2012), so I won't go into any great detail here as to various sightings, apart from stating the fact, that there have been hundreds of them all over the country. An interesting aspect which I haven't covered before is the seemingly close relationship between the lindorm - or at least some of them – and the basilisk. Lindorms sometimes start their life as a tiny worm inside an egg (probably a parasitic nematode worm), and if that worm isn't killed instantly, it will seek out a

dark and damp place to grow, and grow and grow until it reaches full dragon size, but should you chance upon it before it is fully grown, it will stare at you with eyes so terrible, you will either die instantly or turn to stone – accounts vary.

The year of the giant snakes

If we take a step down from the realm of dragons, we get to the world of more manageable reptiles, at least sizewise, although they are still far too big for comfort, and for the Danish landscape. For some reason 1943 was an exceptionally good year for sightings of unusual snakes, and this mind you, was many years before it became fashionable to keep large snakes as pets. During the summer there was a veritable flood of sightings.

In the Klinteskoven Forest on the Island of Møn in southeastern Denmark, there were several sightings of a large boa constrictor like snake, although some people compared it to a giant eel. Several people had seen it when out walking their dog or just passing through the area, and one even claimed his dog had been terrified at the sight of the large reptile judged to be about 3 metres in length.

At roughly the same time, people in Langå in western Denmark was complaining about being attacked or harassed by a large snake every time they crossed a particular bridge just outside of town. This snake mind you, was nothing like the heavy constrictor-thing in eastern Denmark. This was lively and agile animal capable of in one instance curling itself around one poor man's walking stick when he tried to fend it off.

In southern Denmark near the Danish/German border outside the city of Aabenraa, other people had on several occasions seen a big black snake, capable of lifting its head up to the level of a grown man's eyes when provoked, something only big king cobras should be capable off.

I have absolutely no idea what was going on. Had I been presented with a similar set of sightings today, I would immediately have said escapees, but not some 70 years ago.

A few reporters did suggest something along similar lines – menageries, upturned circus-cars and so on, but offered nothing in the way of actual evidence. One intriguing possibility is that all these sightings were in fact euphemisms for the at the time ongoing nazi-occupation of Denmark, but I have talked to the descendants of at least one of the snake witnesses, and they were adamant, that their grandfather had in fact seen and fought a snake during the war.

Some very large lizards – or something

Snakes are not the only oversized reptiles in Denmark. In opposite ends of the country – in the Almindingen Forest on the Island of Bornholm in the extreme east, and in Rold Forest in the north, there seems to be two separate populations of some sort of lizards, much larger than the official lizards of Denmark and much more along the lines of tropical monitor lizards and suchlike. Oh yes – and a tatzelwurm as well!

The Almindingen animals

The Almindingen Forest is centrally located on the Island of Bornholm. It is a big place, so there is plenty to see and explore. In summer lots of tourist pass through it, but most of them only stop near the highest point called Rytterknægten, where a tower gives you an amazing view of most of the island. I can heartily recommend a visit. But should you venture a little bit further afield, there is a good chance of seeing something a bit more special – free-roaming bison is one thing, but for a cryptozoologist there is also some strange almost dragonlike creatures that have been seen several times in the denser parts of the forests, but also close to a couple of marshy areas called Bastemosen and Ølene. De have been described as animal 50-75 cm in length, of a dirty greenish-grey colour, with large eyes, and a small but clearly visible serrated comb along the back and top of the tail. The only living things I can think of, that looks remotely like this, is the marine iguanas on the Galapagos Islands

and the tuataras from new Zealand, and it is kind of hard to imagine how individuals of either species should have found their way to a modest Danish island. But what are they? For now I have no idea.

Although these lizard-like animals are rather special, the by far strangest animal from Almindingen must be a thick set salamander-like animal, which at first glance looks like a very large and fat slow worm, an animal that are fairly common on Bornholm. Upon closer inspection the animal turns out to have small, thin and rather spindly legs, a powerful body and a short fat tail. I have only been able to find 3 sightings of this thing, two from the 1920's, and one from 1974, and although they are quite detailed, I have no idea what it is, although it is very similar to stories from Central Europe about the tatzelwurms, which in fact does not say much, as nobody have any idea as to what they are, although some have suggested that the whole tatzelwurm thing is in fact a hoax. The only known photo of a tatzelwurm has been shown to be a hoax, so who knows.

The lizards of Rold

"I couldn't believe my eyes. It was sitting on a tree-trunk, and it seemed absolutely enormous. For one short moment I actually thought of a dinosaur, although I knew perfectly well it was something else. It looked like a very large lizard. It would have been about 75 cm in length, with a large squarish head, small and very bright eyes, and a distinct serrated comb along the back. I tried to get closer, but it saw me almost immediately, ran down the trunk and disappeared in the thicket on the other side."

There would have been absolutely nothing strange about this story of a meeting with a monitor or iguana or whatever it was, if the story had been set somewhere in the tropics. In t his case though, the eyewitness was in the vicinity of Lake Teglsø in Rold Forest in northern Jutland, and that makes it rather unusual. If this had been the only story of this kind, you would be tempted to credit the witness with an overactive imagination, but this is far from the only time animals like this have been seen in Rold Forest. Since 1967 at least 16 different people have seen one of these animals in different places in the forest, usually close to water, and especially on warm days, where they have been sunning themselves on tree trunks, rocks or in one case, in the middle of a pathway.

These animals look very tuatara-like, but I sincerely doubt that's what they are. It is far more likely, that we are in fact dealing with a small colony of some form of tropical large lizard, or perhaps some of the larger south-European lizards, that have been released into the wild (these sightings are all fairly new, so falls well within the timeframe of keeping exotic pets). Animals like that are so surprising to see in Denmark, that all the eyewitnesses have then added a few extra centimetres to them, and a bit of dragon/dinosaur embellishments.

The dragonstone
To finish the story of the dragons and other reptiles in Denmark, I think it is suitable to tell you a little something about a dragon and a historical mystery all rolled or rather carved into one. And for this we have to go to western Denmark, to the little town of Jelling in Eastern Jutland. This is where we find the two world famous Jelling-stones – one small, one large.

The large Jelling-stone is not only Denmark's, but one of the world's most impressive runestones. It is of such historical and archaeological importance it has been on the UNESCO World Heritage List since 1994. It is a big, grey, triangular boulder, richly carved with runes and decoration. The greyness is caused by the ravage of time. When it was raised, probably sometime in the last half of the 900's, it was painted in strong and dare I say it, rather garish colours. The inscription is quite simple, it is the Danish king Harald Bluetooth stating that the stone is a memorial to his parents, and a place to brag a little bit about the fact that he has conquered Denmark and Norway

and converted everybody to Christianity. Most of the inscription is located on the smallest face of the stone, but it continues onto the two larger faces, each with a carved picture. On one side is a figure of Christ surrounded by a complicated pattern of lines, no mysteries there. On the other side is gets rather strange. Here you can see a strange dragonlike creature looking like it is caught in snares or has been put in chains.

You could of course se this strange being as a symbol of the heathen gods, or perhaps the Devil, having been caught and bound by the new religion. That would all have been well and good if it wasn't for something you can see if you visit the Pergamon Museum in Berlin.

The Pergamon Museum has one of the world's largest collections of art from Asia Minor. One of the stars of the collection is the Ishtar-gate, which is something like 2600 years old. This enormous and richly decorated gate was once a part of the wall surrounding ancient Babylon.

The reconstruction of the gate at the museum shows it being covered in thick glazed tiles, showing various golden images on a deep blue background. There are some extremely realistic depictions of bulls, and some also rather realistic description of a weird dragon-like animal, the sirrush. These are animals with a long scaly body and neck and a distinct mane. They have a large curved horn on the head, and a long bifurcate tongue. The strangest has got to be the legs. They lack scales, but the frontlegs looks like they would be right at home on a lion, and the backlegs look like they have come from a giant bird of prey.

And if you look closely – there is an uncanny similarity to the animal on the large Jelling-stone. We know the Vikings were a busy lot, and that they travelled far and wide – but all the way to ancient Babylon?

Mysteries comes in small packages
According to newspaper reporters and other lower lifeforms, cryptozoology is nothing more, nothing less than the hunt for monsters. It is so much more fun and interesting – and it sells more newspapers - to read about people looking for lakemonsters and abominable snowmen. And it does usually make for entertaining reading, but cryptozoology is so much more, and as I hope to demonstrate in the final part of this chapter, as a cryptozoologist, you can equally well study small and apparantly rather insignificant animals, sometimes just outside your own backdoor, and still walk around with your head held high.

After all, it was our founding father Bernard Heuvelmans who said, that

cryptozoology was the study of unexpected animals – he didn't put any size criteria on them.

So – onwards and downwards. The good thing is that the number of small animals is so much larger than the number of large animals. There are lots of interesting critters waiting out there.

One of the most fertile hunting-grounds for the microcryptozoologist is the storerooms of museums. You would be surprised (you should be) if I told you how much material museums have lying about behind the scenes. Boxes upon boxes of things that have never been sorted or have lost their labels or that nobody have had time to look at.

And there are lots and lots of animals, where all we have is one single specimen, and absolutely no knowledge of how the animal lives in the outside world. In a lot of cases, it has never been seen or studied apart from the specimen lying in a drawer somewhere. Here's a couple of examples just to give you an idea of what I am going on about.

The lonely wolfspider
In 1883 a single female wolf-spider was found on a slope covered in heather in an area in Eastern Jutland called Mols Bjerge. Today this area is a national park, and an area of the outmost scientific interest. The spider was examined at the Natural History Museum in Copenhagen, and lo and behold, it turned out to be a completely new species, which in 1904 was given the name *Pardosa danica*. The animal can still be found in the collection in Copenhagen should you be interested. But here's the thing. Despite the fact that scientists have now being going through Mols Bjerge with a fine toothed comb for more than 100 years, not a single additional specimen of *Pardosa danica* has ever been found. Consequently we know absolutely

nothing about the animal apart from the way it looks. It could conceivably be extinct – the area where it was found originally has changed dramatically in the last century. The specimen found could also have come from somewhere distant, and transported to Mols Bjerge somehow – we simply don't know. But here's a Tasmanian tiger of the spiderworld to go looking for,if you should feel like it.

The dogloving fly with the ridiculous head

In 1794 a rather strange looking fly was found on the carcass of a dead dog in Mannheim in Germany. It was a quite remarkable animal. It was fairly big, dark shiny blue, and with a large round and bright orange head. Local people, who were well aquainted with the fly, even claimed that the orange head glowed in the dark. It received the rather cumbersome scientific name *Thyreophora cynophila* – cynophila means dog-lover. Later it was found in other areas of Germany, as well as parts of France and Austria. And then it disappeared! It was last seen in the 1840's in the outskirts of Paris, and for a long time it was considered to be the first insect to officially go extinct in modern times. The main reason was assumed to be a lack of big dead animals with broken bones to breed on. Apparently the flies laid their eggs in the bone marrow.

And for the next 170 odd years, nothing happened.

But in 2009 the fly decided enough is enough. It was found alive and kicking, or at least

Musca cynophila Mihi.

J. St. fec.

127

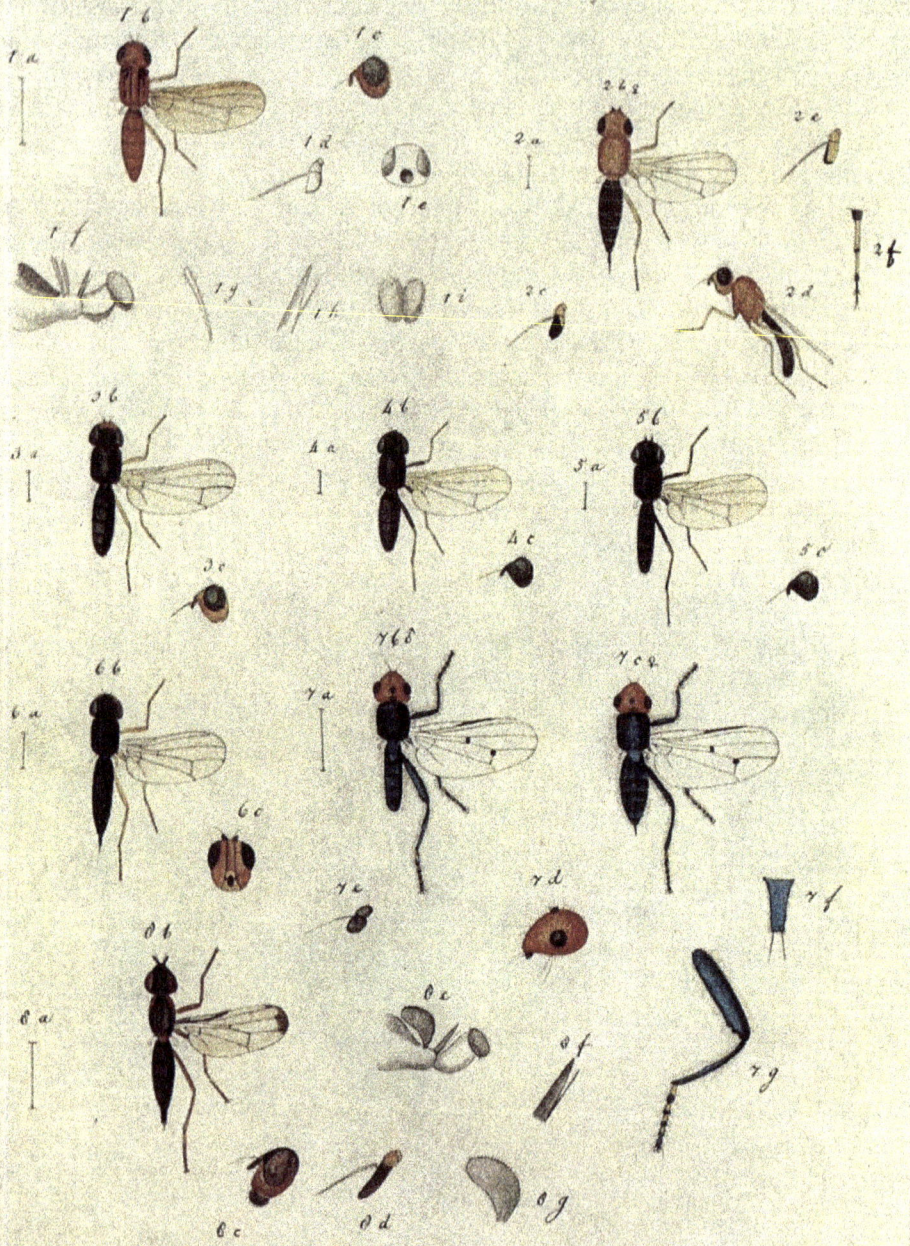

buzzing, in Spain. Of course it had been there all along, it was just that nobody had looked for it at the time where it was most active, and that turned out to be at night in the wintertime.

But the story doesn't end there. Since about 2010/11 I have received several sightings of a large dark fly with an orange head seen in the area around the border between Denmark and Germany. Nobody has so far been able to take a photo or capture one of those flies, but I am working on it. I have tried to arrange for dead animals to be laid out in the area during winter as bait, but so far with no success. Because of the general global warming, many southern insects have started to move into Denmark, and the dogloving fly may be trying to do the same, but haven't been able to get a foothold as of yet, as our winters are a bit more rough than the Spanish ones. We can only wait and see.

Those big hairy things
Denmark, Sweden and Norway, and to a lesser extent Finland hasn't got anything along the lines of the yeti or bigfoot and similar big hairy apelike humans or humanlike apes, but they do have a very large and varied folklore concerning elves, fairies and trolls. There are only minor differences in the stories told in the different countries, so in order to avoid repeating myself to the point of dreariness they will all be dealt with in the next chapter about Norway. This seems only fitting, as Norway is by far the country with the most sightings of these creatures, and by far the country with the most living tradition of trolls.

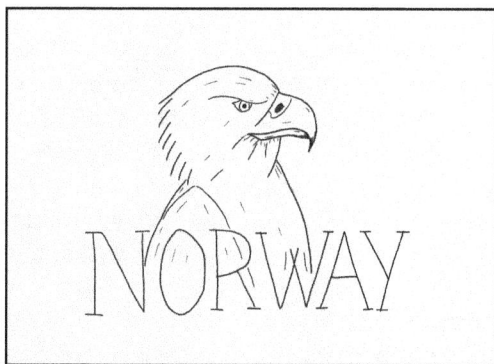

NORWAY - Mountains, mountains and yet more mountains

Imagine a country shaped like a giant tadpole – throw in a chain of mountains – and you've got Norway. The southern part of the country is a big, almost circular rocky plateau where the capital of Oslo and most of the larger cities are located. The tail end of the tadpole is a long narrow strip of mountains stretching far towards the north, way beyond the Arctic Circle.

Rocks and mountains are the main things in Norway, but even though the population of Norway is fairly small, it is far more difficult for large animals to hide themselves in the Norwegian mountains. The Norwegian people are avid skiers – most of them from an early age – and the mountains are not tall enough, steep enough and remote enough to keep the number of visitors down. But still, there are creatures here, strange cats, big birds and suchlike – and there are trolls!

Land of trolls
If ever there was a land made for trolls, it is Norway, with its mountains, fiords and fantastically convoluted coastline. It is a breathtakingly beautiful country. And it is packed with trolls. They abound in Norwegian life and folklore. They are also common in Sweden and Denmark, but because they are so prolific in Norway, I will treat them all here.

Before we proceed, at word of warning might be in order. It can be a bit difficult to gather sightings and stories of weird animals in Norway. Not because nobody has seen anything, but because the locals – even today – are strangely reluctant to talk about the creatures. Perhaps they fear ridicule or something similar, but I have found you do need to work quite hard to get them talking. But with a little persistence – and a drink or two – you can usually get some of them to open up. It's not cheap, mind you. The price levels are frighteningly high in Norway, but on the other hand, the Norwegians are fairly effluent thanks to all the oil from the North Sea – but never the less, you might have to save up for a while, before you go to Norway. The first time I tried to check into a youth hostel in Oslo, I had to pay what a room in a decent hotel would cost in Denmark – that little fact shortened the vacation with at least a week.

The natural history of trolls

Trolls are without doubt the quintessential legendary creatures. They play parts – speaking and otherwise – in countless tales and other stories. They can be found in books, plays and movies. There is no question about the fact that trolls walk around dressed in extremely heavy fur-coats of legend, but if you try to undress them so to speak, indeed if you try to sift your way back in time through the hundreds if not thousands of troll-stories there is, you get a growing sense of reading about actual flesh and blood creatures. The older the stories, the more information you can find of their daily life, and, dare I say, the biology of the trolls.

Some researchers have even suggested that the Scandinavian stories about trolls are the equivalent of the stories about Bigfoot, Sasquatch, almas and orang pendeks you hear from other parts of the world. There are indeed great similarities between f. inst, the Danish Slattenpatte (Floppy Tits) or the Norwegian Slattenlangpatte (Long Floppy Tits), so named because the females have very long and pendulous breasts, and some of the various abominable creatures. Some of the Swedish and Norwegian trolls in fact have so long breasts that they sometimes trip in them and fall over. Trolls also have a reputation of smelling quite awful. Something you also often hear repeated in eyewitness sightings of big hairy monsters – the most extreme example of course being the American Skunk Ape. The smell is actually a bit more subdued in trolls – they either smell like wet, damp and rather dank earth, or they smell like they haven't washed for a couple of hundred years – which they probably haven't.

The classic image of a troll is a large, sometimes very large, hairy and smelly humanlike creature. A troll typically has a very big nose, big hands and feet, sometimes, although not always and occasionally even horns. Trolls are generally ill-tempered and grumpy. They can sometimes speak, but in some cases they do nothing but roar and grunt. They don't like loud noises, especially church-bells, and will do quite a lot to destroy churches.

The bells, the bloody bells!
In the central part of Denmark, just outside the city of Nyborg, you can, if you follow

the signs, come upon The Ladies Rock or Dammestenen, the biggest single rock in Denmark. According to legend it was thrown by a troll who didn't like the sound of the church bells in Nyborg. It fell several miles short of the church, so the trolls aim wasn't too good. But it was a rather big troll at can be seen by the fact, that I am standing next to the stone on the photo, and the stone is variously described as being a pebble found in the trolls pocket or apron.

Another angry troll left his mark – a footprint on the wall of Skamby Church, also in central Denmark. He tried in vain to kick down the walls of the church during the building of it, but unfortunately for him, the church walls had already been laid out in the shape of a cross, so it was impossible for him to do any lasting damage.

Suns and lovers

Trolls don't like the sunlight much either, and some of them have an unfortunate tendency to catch humans and eat them. Sometimes they abduct humans without harming them – instead they use them for much more nefarious purposes. Big old male trolls are very fond of shapely young girls, and female trolls like strong young men as well. And what do they want them for? Breeding! Apparently trolls are rather

inbred, so they need regular infusions of new blood. Human that the trolls take as lovers are generally treated very well and there are stories of people trying to free other people who had been taken by trolls, only to be met with a blank refusal. Trolls seem to be very good partners and lovers when they need to.

From a biological point of view, this would indicate that trolls and humans are in fact closely related. But if you dig deeper into the stories you find that the case is in fact far more complicated. There are for instance as I have mentioned before, several different types, dare I say species of trolls.

Trolls have a rather large distribution in Northern Europe. They are most common in the Scandinavian countries – Denmark, Sweden and Norway, but they also spill over into northern Germany and eastwards into Finland. Up north in the Arctic, trolls are extremely rare – they quite simply don't like to be cold. They have also followed the Vikings to Iceland and the Faroe Islands, and even Greenland has their fair share of stories about troll-like beings – more about which can be read in the appropriate chapters of this book.

Trolls are not especially social creatures. Old males, especially in the big species, usually live alone. The smaller species live in small family groups – a male and a female and a variable number of youngsters. They typically live in some sort of cave or hollow in the ground. According to Scandinavian legend most of the burial mounds we still find are in fact old abandoned troll dwellings.

Not much is known about the breeding of trolls, but as they can get quite old, and there are stories about the same troll living in a particular area for several human generations, they probably only have a small number of young, and said young take a long time to grow up. By and large trolls take things fairly easy. They sleep a lot, and some of them can take a month to wake up properly. They are at least 10 years old before they can walk, and 20 years old before they can talk. There are also information about humans (especially children) making friends with young trolls that turn out to be maybe 50 or 60 years old.

But apparently trolls are not especially fertile – stories abound about trolls abducting healthy human babies and leaving sickly troll babies instead. Abductions like these – as well as the troll's general aggressive and destructive behavior towards churches and suchlike, have led to some animosities between humans and trolls, but in general they have left each other well alone. In actual fact trolls can be good and faithful friends if you treat them with respect and compassion. One has to be careful around trolls though – especially the Norwegian trolls, as they have a decidedly nasty streak. Swedish trolls are a bit more shy and withdrawn, and some of the Danish trolls are full-fledged cowards.

But even friendly trolls can have a hard time controlling their mischievous streak. Parykmanden – or The Wig Man – is a small Danish troll. He is friendly and generally merry, and has been named because of his long and rather luxuriant but shaggy hair. He likes to roam the more isolated country roads of northern Denmark and put the wind up any fair maiden that wanders around after dark. He especially likes to sneak up on them and peer under their skirts or have a little feel!

Night and day
Generally trolls are considered to be nocturnal, but in fact most trolls are active during the day. They just keep to themselves if they can – and they are very good at hiding. Although in some areas, the authorities have found it necessary to warn people about them. In the Middle Ages warnings like this was taken very seriously indeed, but today, especially in Norway, it is very much exploited by the tourist industry.

The nocturnal trolls are perhaps best known for turning into rocks on exposure to sunlight. Stories like The Hobbit and several others have made this phenomenon very famous. Some nocturnal trolls are a bit different. They turn into trees, when they feel the rays of the sun. Strangely enough this ability varies depending on where the troll lives. Mountain trolls which are the most common type in Norway, always turn into stone, whereas only forest trolls, and that's mainly the Swedish ones, turn into trees. A few very old stories describes ice- or snow trolls living in the very far north of Scandinavia, and they turn into ice- or snowdrifts if they see the light of day.

Now skeptics will say this belief is based on sightings of humanlike rock formations, trees or ice chunks.
But from a biological point of view, it might as well be a highly advanced form of camouflage. After all camouflage is well spread in the animal kingdom, and some animals are quite incredibly good at looking like rocks, tree stumps or other inanimate objects.

Trolls of a different size
Finally I thought I would give you a short rundown of the various known species or perhaps I should just call them types of trolls in the Scandinavian area. There are in fact quite a number of them, but they can be separated in three different families. Some of them are similar to the trolls found in Iceland, and it is possible they are in fact the same.

The first family is the small trolls. They are all smaller or of the same size as humans. They are not especially hairy, often lack a tail and can be mistaken for humans. They also seem quite closely related to humans as the two species are capable of interbreeding. The small trolls are diurnal, but usually keep to themselves.

Judging from the descriptions of the small trolls, there are at least six different species. They are most common towards the south, especially in Denmark and southern Sweden.

The medium sized trolls are the classical trolls of the folklore and fairytales. They are from the size of a normal human male up to a height of about three metres. They have a very broad and solid body shape. They are very powerful and hairy, and usually have a tail. They have a big nose and quite coarse features. They can to a certain extent interbreed with humans – but often they abduct human children and replace them with their own. They probably suffer from quite a lot of inbreeding related problems. According to the old descriptions trolls have never been common, so it is only natural to assume that inbreeding has become more and more an issue as time has gone by. Judging from the descriptions of the medium sized trolls, there are at least three different species, possible four: two in Denmark, one in Sweden and a possible one in Norway.

The final group is the giant trolls – very large creatures, up to 10-20 metres in height, i.e. the size of various smaller dinosaurs, and with what one must assume is an equivalent weight. Giant trolls are all nocturnal, and they are all susceptible to turning into stone or trees. They look very much like their smaller relatives, the medium sized trolls, but they can have superfluous bodyparts, several heads, extra arms and so on. Some troll researchers have suggested that they are a sign of great age and status, somewhat similar to the growing size of antlers in a deer. Giant trolls are by far the rarest of the lot. Not much is known about them, as people have usually only seem them at a distance, wandering in the mountains or in the woods at night, or in their petrified or lignified forms. They are most common in the mountains of Norway and Sweden and in the very far north of Scandinavia.

Despite their various differences, the giant trolls are probably just one or possibly two different species, as some of them seem only to turn into stones and others only into trees. The type that turned into ice have not been seen for centuries and is probably extinct.

Now, to finish this off, I could put up a diagram showing the relationship between humans and the various kinds of trolls, I could also be very skeptic and claim it has all been a figment of peoples imagination, a way of giving the dark and unknown some sort of recognizable form. But, although a lot of what I have told you and shown you are of course quoted from fairy tales and suchlike, I cannot help but wonder, if somewhere buried deep inside the whole mass of troll stories are in fact sightings of real living creatures made thousands of years ago. Perhaps bears, perhaps other people seen from afar, perhaps mirages, heiligenschein or something similar, and perhaps, just perhaps, distant memories of a time when the modern human, Homo sapiens sapiens, wasn't alone.

I am fully aware of the fact that the Norwegian movie The Troll Hunter (2010) is sort of a comedy, but nevertheless, it is worth watching, because whoever wrote the script filled it with lots of facts about troll-biology taken from the many legends. You can do far worse, than spend a couple of hours watching that. And the special effects are remarkably good.

4 sightings and a famous legend
Finally I will present 4 sightings of trolls – two from Norway, one from Denmark and one from Sweden to give an idea of what we are dealing with when we go out and about up north. And no book of mysterious beings would of course be complete without the story of Beowulf and his fight with Grendel and his mother – a fearsome troll if ever there was one.

Going home
The Trysil area is one of the most popular and most visited skiing areas in Norway. It is located some 150 km northeast of Oslo, along the border with Sweden, and today it is an area of hotels, huts and a multitude of skies and skiers. If you go there in the summer, it is also a good area for walking, and fairly peaceful as well. The area covers some 3000 square kilometres, but even today the total permanent population is only around 7000 people. There has been quite a number of troll sightings in this area going all the way back to the 15- and 1600's, but also a couple of fairly recent ones.

In the summer of 1972, a 22-year old woman from Oslo on vacation was walking along Revhibekken (Revhi Creek) towards the little town of Ljördalen. It was late afternoon, and the weather was rather wet and misty, so she was quite anxious not to lose her bearings. She was as a matter of fact so anxious, that she focused her entire attention on her compass and forgot to see where she was going and walked straight into what she for a few seconds thought was another walker coming the other way – but this was a kind of walker she has never encountered before.

> *"Whoever it was, was so big and powerful, that he or maybe it, knocked me straight on my back, and I was carrying a rather heavy backpack at the time, so that was no mean feat. As I lay there like a capsized boat, I started yelling curses and asking for help to get up again, but all I heard was a deep almost coughing grunt, sounding something like "Nei"* [the Norwegian word for no], *and then I got a whiff of something*

that smelled like a very old and extremely disgusting locker-room. I was so angry I kept yelling and cursing and trying to get out of the straps of my pack, but by the time I'd managed to free myself and stand up, the thing I had collided with had walked past me, and was now maybe 50 metres away. Only then did I realize, that the other walker was not human, it couldn't possibly have been. It was tall – maybe 2½ metre, with a massive head that merged almost straight into a massive furry back and two short, thick and stumpy legs. My mind went completely blank – but for some reason I reacted by yelling "Watch where you're going you clumsy oaf!" at the top of my voice. That actually made the thing stop, do a half-turn, and look at me over its right shoulder, before it gave another deep grunt and walked on. I am not totally sure, but when it turned, I think a saw the glint of a small eye and possibly a big nose or maybe a horn, at least something protruding from its head."

The trouble with tires

I am not entirely sure of the next story, the reason being I have heard it in several different versions, or rather, the story is the same, but the actual point in time varies with almost 20 years, so it may be some sort of urban troll legend. Never mind, it is still a good story. It concerns a man from Oslo, who were driving north towards Gjøvik, at town located about 100 kilometres north of Oslo. As luck, or unluck would have it, about 10 kilometres outside of Gjøvik, one of the tires on his car decided it was time to abandon ship and have a puncture. It was autumn, it was late, the light was fading, and he had to change tires as quickly as he could. There was no traffic on the road, so he set to work as quickly as possible, but all the time had the strange feeling that he was being watched. On several occasions he thought he could hear something breathing, and when he dropped his tire-iron, he was certain he heard a sharp intake of breath. Nothing untoward happened, but when he put all his gear, and his flat tire in the back of his car, he chanced to glance towards a row of trees to the left of the road, and some 2½ or perhaps 3 metres above the ground, he saw a very large yellowish eye staring at him. At that point the man decided that an evening meal in Gjøvik and a very large drink was just the thing, so he vacated the place as quickly as he could.

A very large creepy crawly

Some years ago, when I was in Sweden, or more specifically Storsjön, doing my best to rub the natives up the wrong way (I kept insisting that most of the sightings of the Storsjö-monster was sightings of swimming moose, and most of the locals didn't take kindly to that idea), I met a guy who was certain he had seen something completely out of this world when he was a small boy in the late 1960's. He had been out very early one summer morning with the idea of doing a spot of fishing, but he never got around to it, because while he was making his way towards the lake (to a place where the game warden wouldn't see him fishing), he saw something really strange near the lake.

> *"I clearly remember stopping and looking ahead, and saying out loud: "That's funny – a tree with fur!" It basically looked like a very large tree-trunk lying on the forest floor covered in something that looked like moose-fur or beaver-skin or something like that. Then I started wondering whether the trunk was covered in wilted moss, and then I started to get nervous, because it dawned on me, that the tree or whatever it was, was actually breathing and grumbling like my father did when he was taking an afternoon nap on the sofa every Sunday. Although I was scared, I kept walking closer, but forgot to look where I was going and stepped on a dry twig that broke with a loud snap. In the next second the brown thing woke up, but instead of getting up and running away, or for that matter attack me, which I was sure it was going to – I actually wet my pants from fear – it seemed to grow four hairy limbs within seconds, and started crawling away at high speed. It passed fairly close by me, and I could see it was shaped like a very large and hairy human. I don't k now why it didn't just stand up, but it didn't. It just went on, and then it disappeared among the trees and was gone.*

The troll-hunter – Danish style

For the last troll history we must turn towards Denmark, and to give an idea of the scope and width of troll stories, this one I have taken from the land of legends, although there is some tantalizing evidence of it not being entirely imaginary. The story is about an old Danish king, trolls and ghosts – so what more could one ask for in a single story?

According to the history books, the Danish king Valdemar 4 (1320-1375) came to a sticky end. He was so fond of his castle at Gurre, about 40 kilometres north of Copenhagen, that he allegedly said that God could keep Paradise to himself, if only Valdemar could keep Gurre. God didn't like talk like that, not even from a king, so he decided that Valdemar should ride the roads of Denmark until Judgement Day, always with a flock of baying hounds, and always hunting, devils, demons – and trolls. And he gets around the old king, because the Vejlø Bridge south of the town of Næstved, itself about 60 kilometres south of Copenhagen, is part of the kings hunting ground.

According to an old story from the 17th century, late one night a man was walking

across the Vejlø Bridge, when he suddenly saw a giant woman running towards him. She was naked, with long pendulous breasts that almost reached to the ground, and she was running like mad, gasping and wheezing. She disappeared as fast as her fat legs could carry her into the darkness on the other side of the bridge. A moment of silence followed, and then three enormous hounds ran past the man. Their eyes were gleaming with hellish fire, and flames were spewing from their gaping maws. They looked absolutely terrifying, but they ran past the poor man without noticing him at all.

Now the man knew what was going on. King Valdemar was on a hunt! And who should appear a few moments later but the king himself on a white charger. The king brought the horse to a stop, and politely asked they man whether he by any chance had seen a Floppytits passing by. The man told the king about the giant woman he had seen, the king thanked him, and rode on. A few minutes later the man heard a horrendous scream coming out of the darkness on the other side of the bridge, and he realized that the king had found his prey.

So much for the legend but what about the evidence? Well, should you be so inclined, you could for instance pay a visit to Vejlø Church, where a wooden effigy of a Floppytits is actually holding up the pulpit. The statue is several hundred years old, evidence of the age of the story, but locally people also tell a story about how a set of tracks of extremely large humanlike feet was found close to the bridge in 1886. Unfortunately I have never been able to dig up any precise details about that particular case, but it is quite intriguing.

And the king? Well, he is still out on his hunt. If you stop on the Vejlø Bridge on a cold and frosty winters night, you can sometimes hear the hounds and the sound of a

horse galloping past. Once upon a time, you could actually see the king, but perhaps the many years of hunting has worn him out, so only the sound remains. That on the other hand has been heard as late as 1982 by a nocturnal cyclist.

Beowulf, Grendel, Grendels Mom – and a dragon for good measure

The ancient legend of Beowulf tells the story of a brave warrior who sometime in the 6th century travels from his home in what is now Sweden, to help a Danish king who lives close to what is now the town of Lejre in Eastern Denmark. The Danish king are having an awful lot of trouble with Grendel, a giant troll-like monster, who has developed a nasty habit of breaking into the kings great hall at night, and killing and eating his men.

Beowulf, being a true hero, knows exactly what to do. He pretends to go to sleep in the great hall, and when Grendel returns, a mighty fight starts, with Beowulf fighting the troll with his bare hands. Although the troll is extremely strong, Beowulf wins the fight by tearing one of Grendels arms clean off. The monster – deadly wounded - runs away to its lair in the nearby swamps. Everybody of course rejoices by the fact that Grendel will not bother them anymore, but all is still not well, because even monsters have mothers…

And Grendels mother is as bad, fearsome, terrible and dangerous as they come. She is furious with the men for having killed her son, so the next night she turns up to have her revenge. The ensuing fight is even more violent and dramatic than the first one, and Beowulf ends up having to follow Grendels mother into her cave beneath a lake. Here he finally ends up victorious by taking a sword from the hoard of treasure in the cave, and chopping off her ugly head.

Everybody salutes the hero, and there is much celebration and feasting before Beowulf heads home to Sweden to set up his own kingdom. Fifty years later he even fights a dragon, by the way, but that's another story.

I am not for a second suggesting this is based on a true story as they say in movies and the seedier types of novels, but it in rather thought provoking, that stories like that have been told for such a long time.

A cryptozoological roadtrip

After this rather lengthy tale about trolls, I feel it is about time we turned our attention to something completely different. So just for a change let's go to the central Station in Oslo, and get onto a northbound train towards the Atlantic Coast. The trains are rather comfortable, and for the first part of the journey, you go through the rather nice and leafy suburbs of Oslo, but then the line starts to climb, and suddenly you realize you are driving through two of the most legendary mountain areas of Norway. To the west is Jotunheimen, with the highest peaks in Norway, and to the northeast is Dovrefjeld, home of some of the biggest and meanest trolls in Norway. The famous Norwegian composer Edvard Grieg even wrote a piece of a visit to the hall of the mountain king, or Dovregubben, as he is known in Norway. This is more or less the signal to sit with your face glued to the window, because, as a woman sitting on a seat across from me told me on my first trip to Trondheim – "you never know what you see on Dovrefjeld". I knew you could see musk oxen with a bit of luck, but there are apparently other strange creatures living out there in the wilderness as well.

My traveling companion turned out to be a veritable goldmine of information. She had lived in the area most of her life, and had spent countless hours out and about, and had seen some weird things in her time.

The demon bear

"When I was a young girl, my parents told me stories about a demon bear that you could meet, if you got lost in the mountains. This bear was very light in colour, almost white, and could look like a polarbear, although you would never normally see polarbear that far south. But it was not a polarbear – it could run like the wind, and it had very large ears, so if you ever saw one, you had to keep completely still. I didn't really believe my parents, but one day, when we were driving by car through Jotunheimen – we had been visiting relatives in Sogndal and were driving towards Lom – maybe I was 15 or 16 years, and I was very bored in the car. I had turned around in the backseat and was looking back the way we had come, when an enormous bear suddenly ran across the road. It was in the middle of the day, and the weather was fine, so I saw it clearly. It was absolutely huge, and it cleared the road in two jumps. I had never seen anything like it, and I have never seen it again, but I am sure it was a demon bear."

Now there is a population of bears in Scandinavia, but these are common brown bears, and like all other bears they are brown, with rather short legs, and although they can move quite fast, they look nothing like the animal my train-friend had seen. She also insisted that it didn't move like a normal bear, which she had seen on several occasions. It had longer legs, and was indeed white. It is highly unlikely, although not completely impossible, that a polarbear would be able to make its way down south from the Arctic – there is in fact and old story from Denmark about a white bear suddenly appearing on the coast close to the town of Århus during a very harsh winter - but although they are white, you cannot with even the best intentions describe them as long-legged. In some areas of Western North America, you can also find white of very light brown bears – in some areas known as ghost bears – but they are not long-legged either – so what on earth did she see? She could of course have been imagining the whole thing – she was after all a teenager bored out of her skull – and as such fully capable of winding her parents quite severely up. But when I talked to her, some 30 years after the occurrence, I had no sense of the story being a fabrication.

If the animal was real, the only possibility I can think of, is an animal similar to the strange bear which have been seen in places are far from each other as Sibiria and Northern Greenland – a strange long-legged animal with a short face and a weird way of moving. This might be surviving individuals of more primitive and ancient species of bears, but it is by and large only guesswork.

"By the way," my newfound friend said, as she suddenly rose to get off at the next station, *"remember to look out for the wolves and the jackals as well!"*

White wolves and blue jackals
Now wolves I could understand. Norway does after all have a small wolf-population, although it is a highly controversial subject in Norway, and you can quite quickly

alter the mood of a room from convivial to frosty by saying something positive about wolves. But what did she mean by jackals? The country has a good population of arctic foxes, but they are rather small animals, and all though they have a lot of personality, I found it hard to imagine them being compared to jackals.

It actually took a bit of research before I discovered that stories about weird wolves and blue jackals are fairly common in these areas, especially around Dovrefjeld, and that they actually go back to at least the middle of the 1500's, when bishop Olaus Magnus wrote about them in the book about the Nordic People.

The stories about the wolves are fairly straight forward, and as such there is nothing mysterious about them. Magnus writes, that in Dovrefjeld you can see white wolves, and that their main prey is smaller and weaker animals. Well, 500 years ago, the wolf-population was no doubt considerably bigger than it is now, and we know from f.inst. Greenland and northern Canada that wolves sometimes are white, and it is easy to see that perhaps an isolated population would have a rather large proportion of white animals, as this special colouring probably is genetically caused.

Things get a bit more strange if you read more off Magnus's description. Because he also writes that:

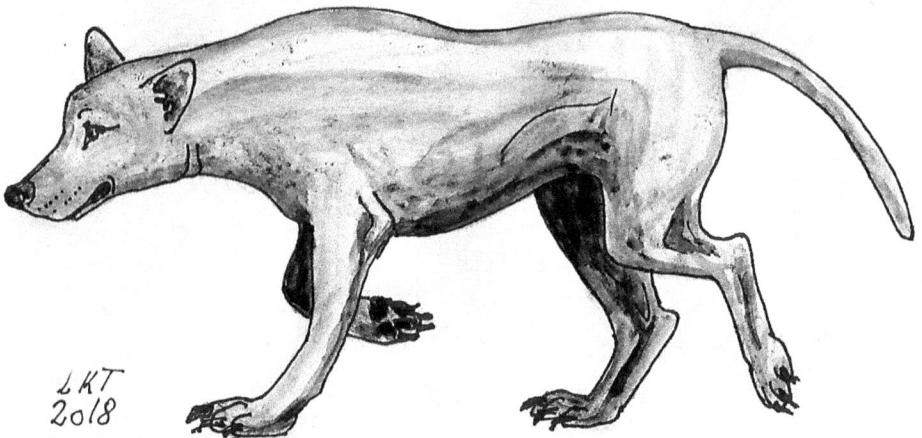

"..there is also a kind of wolves called jackals, with a more elongated bodyshape. They are also different because they have shorter legs. They are quick and live by hunting. They are furry in winter, but naked in summer."

Normally I would have had no trouble in ascribing the jackals to arctic foxes, because they are very much furry in winter and look almost naked in summer – they are in fact so different looking that you could be forgiven for thinking them two different species. The only problem is that Magnus describes the foxes as well, so his jackals must be something else. And I have in fact managed to dig up a few sightings of these jackals from a later date – a much later date, although nothing newer than the 1950's.

If you put these sightings together, you get an animal considerably bigger than an arctic fox – something along the line of an Alsatian or a similar sized dog or a small wolf. Most of this description is taken from the story of a man who at the end of the 1950's lived in a house on the outskirts of Lillehammer. He was a boy of 13 or 14 at the time, and for most of the summer of one particular year – he wasn't quite certain when I spoke to him, but thought it would have been 1958 – a very strange doglike thing was lurking in the neighbourhood. He saw it at least 7 or 8 times, and was very intrigued by the fact, that the animal appeared to be completely naked. He couldn't see any kind of fur, and the skin of the thing had a very strange colour that he compared to the bluish-gray colour you find on the back of a wood pigeon. It looked very smooth, and although the animal never seemed particularly afraid of him, it never allowed him any closer than 5 or 6 metres.

Now this sound very much like the American blue dogs, or Elmendorf-beasts or whatever names they go by - a dog-like animal with little or no fur and a bluish-grey skin with no signs of scarring, scabs or anything indication an attack of mange.

It is not very likely, that there would be a completely unknown species of dog-wolf like animal living like that in Norway, but it does rather suggest that the various blue dogs – including the Norwegian version, is in fact caused by some form of genetic mutation, that apparently can happen spontaneously in isolated populations of doglike animals. My money is on the wolves in the area.

The city of dragons
It takes the better part of a day – some 7-8 hours – to get from Oslo to Trondheim by train, and when you get there, you step out into what seems like a completely different world. Trondheim is a beautiful place, but with a bit of a split personality. It is a fair size city, some 200.000 inhabitants, but with the look and feel of a small provincial fishing village with lots of wooden buildings.

The most impressive sight in Trondheim by far is the Nidaros Cathedral. It is absolutely enormous, and it is not only the cathedral of Trondheim, but the official state cathedral of the entire country. This is where the kings and queens of Norway are crowned, and where all kinds of stately weddings and funerals take place. It is also a church with roots deeply embedded in the pagan Viking past of this part of the world. There are carvings of dragons and giant snakes everywhere – the latter are probably depictions of the Midgaardworm or Jormundgandr, the giant snake of Nordic mythology that completely encircled the globe, or Nidhug, the dragon that constantly gnaws the roots of Yggdrasil, the tree of the world.

But giant snakes and enormous dragons are not just a thing of the ancient past in Norway. There are stories from large parts of the country about weird reptilian creatures – some as recent as the latter half of the 20th century - which I suppose is rather strange in a country whose natural reptile fauna comprises only five small species, one lizard, the slow worm and three different smallish snakes.

There are quite a number of sightings and stories of dragons in Norway, or rather giant snakes as it were, because Norwegian dragons are extremely long and worm-like, in some cases described as up to 50 metres in length. They never have wings, and only some of them have legs, but in some cases they are like basilisk capable of killing people just by looking at them. They are quite capable of spitting fire as well, or poison, and they can generally create absolute havoc in a very short time if they put

their mind to it. It is no wonder that several Norwegian legends claim that the dead sometimes rise from their graves to fend off the dragons if their population have risen to plague proportions or they are just being too much of a nuisance.

The oldest description of what is supposedly an actual sighting I have been able to dig up dates from 1522 when the good people of Moss about 50 kilometres south of Oslo were having trouble with one of them. The description of the actual animal is not very detailed, but it did have an unfortunate habit of gobbling up people's sheep and cattle on a fairly regular basis. Several attempts were made to chase the animal away, or indeed kill it, but all to no avail. After each of its destructive excursions it would disappear into the wilderness away from the coast, and one day it simply didn't come back. I suppose the local's collective sigh of relief could be heard in Oslo. In the following centuries the snake-dragons were seen several times. All the sightings were spread out along the coastal areas of southern Norway which might indicate that these animals were in fact aquatic and only occasionally ventured onto dry land, although no one apparently had seen them coming up from the sea or going back into it.

When it comes to the look of the animal, all the sightings are remarkably consistent apart from one minor – but quite important detail, but I'll come to that later on.

The look of it

The animal is distinctly snake-like, with a slim but powerful body covered in small but distinct scales. The underside is white, or at least considerably paler than the topside, which is typically greenish brown with various blotches and irregular stripes. In some cases people have seen a ridge along the back og even an irregular comb-like structure. There is usually some form of mane-like fringe at the back of the elongated almost crocodile-like head, with large eyes and prominent teeth.

Eagle-eyed readers will have noticed, that I haven't written anything about the size of the thing – as according to the sightings it has been constantly shrinking for the last 4 or 500 years, and so has the number of sightings. I have only been able to find two modern sightings of the snake-dragon, although the dragon bit is hardly appropriate anymore.

The first sighting comes from a woman who was out walking her dog just north of the little town of Mandal in the extreme southern part of Norway. It was a sunny spring day in 1968, when she, or rather her dog, chanced upon something neither of them had seen before.

> *"My dog had run on ahead, and had disappeared from view. It was something he often did, so I was not worried, but suddenly he started barking like mad. He became more and more frantic, and although I tried to call him, he didn't react, he just kept on barking. I hurried along to find him, but almost wished I hadn't. He was standing at the foot of a large rockpile, and on the top of that, sunning itself, was the biggest snake I have ever seen. I suppose I should have stayed and taken a closer look, but I was so frightened I just grabbed my dog and ran more or less straight home with him in my arms. I had never seen anything bigger than a grass snake, and this was as thick around as an old fashioned barrel. I have no idea how long it was, it just looked like it could have swallowed me and my dog in a single gulp. I never told anyone, and I waited for weeks to hear if somebody else had seen the thing, but nobody said anything, so I the end I kind of assumed I had dreamt the whole thing – but thinking about it today I am sure it was a real animal."*

I don't really know what to make of this sighting. To me it looks like perhaps an escaped pet snake that this woman's fear made considerably bigger than it actually was. The only thing is that big pythons were not exactly common as pets at that time, and that an escapee would surely have made the newspapers – and I haven't been able to find anything about that at the time. And even if it should turn out to be a python – what about the older sightings? They might in fact be a mixture of legends and sightings of things that might be snakes, but might also be other things that just happens to snakelike, such as this next sighting from 1978 from a wooded area on Bygdøy just outside of Oslo. This was seen by a group of school-kids on a field trip to the Kon Tiki Museum which is also located on Bygdøy.

After their visit to the museum they had gone for a walk with their teacher looking for a suitable place to eat their lunch. They were walking to a forested area when a couple of the kids called the teachers attention to a strange snakelike animal crawling slowly across the road 20-30 metres in front of them.

> *"The whole animal seemed to be pulsating and moving with a strange rippling like a big flat caterpillar. It was the strangest looking snake I have ever seen. I would have liked to go closer, but I had 12 young children with me, so just in case the things was dangerous, we all kept back and just watched. The animal moved quite slowly – it took several minutes to cross the road, but at one stage it was stretching from one side of the road to the other and both ends disappearing into the undergrowth. We measured the road later, and that showed us the snake or whatever it was must have been at least 6 metres. I have simply no idea what it was."*

I might though – it sounds to me like what we in Scandinavia in general call armyworms, but which is in fact a mass-migration of the larvae of various species of fungus gnats. They move like a strange slithering army, and although they can look like one big entity, it is in facts thousands of larvae crawling over one another. Nobody knows why they move like this – the armyworms are formed and dissolved without any apparent cause, and they sometimes only stay together for a few hours. All of which may come a long way in explaining why these weird pulsating "snakes" sometimes seem to vanish into thin air when you look for them – especially if the search doesn't get underway until the next day. Although you may run into a bunch of bloated and burping birds that have spent a lot of time gorging themselves on the larvae, and thus making it even harder to find them again.

Speaking of birds...
If you go to the west coast of Norway, which is fairly easy, as most of the country consists of the west coast, you quite quickly get into the territory of one of the world's most impressive birds of prey, the white-tailed sea-eagle. The Norwegian population of this magnificent bird is one of the biggest in Europe, if not the biggest. And this might well be the reason why there are so many stories about giant birds in Norway –

not that the sea-eagle actually needs any kind of a boost from legends. It is more than impressive enough as it is. I have been an avid bird-watcher since I was 14-years old, and growing up in Scandinavia has made it possible for me to see sea-eagles on a regular basis, and I guarantee that once you have seen an eagle in full flight – it looks suspiciously like a barn door – you are not likely to forget it in a hurry. And if you have ever been really close to one of them – suck as banding young eagles, they make an incredible impression. To say they are huge is an understatement, and seeing their talons up close, is like seeing a real live version of Freddy Krueger from Nightmare on Elm Street. It looks like the bird is holding a couple of fistfuls of steak-knives. It is so easy to ascribe this awe-inspiring creature a lot of abilities that it doesn't actually have – such as incredible strength and the ability to fly off with everything from lambs, full grown sheep, reindeer calves to dogs, cats and young children. I have actually once found bones from a domestic cat under the nest of a golden eagle, but never a sea-eagle. They do in fact prefer birds and fish. And if you come in close contact with the big birds, you will get a surprise, because they are surprisingly light-weight birds.

Even very big females (as in all birds of prey, the females are bigger than the males) only weights 7 or 8 kilograms. Nevertheless, the birds have been fingered as the culprit in some rather gruesome killing and/or abduction cases, and none of these are probably more famous than the one that took place on a clear day in the summer of 1932. Several books have been written about this case, and there has even been a film made. It has been a case of much controversy in Norway, and as such is worth a closer look.

Svanhild and the eagle
On Sunday June 5th 1932 3½ year old Svanhild Hansen had been to church with her family on the island of Leka which is located in a part of Norway called Nord Trøndelag, some 700 kilometres north of Oslo. Following the sermon Svanhild and her family went to a nearby farm to stay with relatives. About half past three in the afternoon Svanhild was playing alone outside, a short distance from the house, when she suddenly disappeared. On that day there as a lot of people in Leka, so within a fairly short space of time, a search had been organised – according to some sources only about 10 minutes after she had gone missing – consisting of close to 200 people. For the first 5 hours not a trace of the girl could be found, but then one of her shoes was spotted lying at the foot of the Hagafjäll (Haga Mountain) on the island, and people started talking about whether Svanhild had been taken by an eagle. There was a local pair of sea-eagles breeding on the mountain, and the birds had been extremely agitated all afternoon. There was no sight of Svanhild, but three local young men decided to try and climb the mountain and search for the girl. It took them almost two hours to get close to the nest of the eagles, and here they found Svanhild on a narrow

ledge, where she had crawled into a hollow in the rock-face – 1700 metres away, and 180 metres above the yard where she had been playing 7 hours earlier.

After her rescue, the little girl was examined by the local doctor. He could find no wounds or damages on her apart from a scratch on her forehead. Her dress and various layers of clothing, apart from the innermost one was holed and torn, presumably done when the eagle had grabbed her. She was missing one of the shoes, but that had already been found. Although he had his doubts, after having talked to the girl and been shown the place where she had been found, the doctor was forced to conclude she had indeed been taken by an eagle, as he could find no other explanation, and because nobody thought she could have walked to the mountain herself.

When interviewed Svanhild at first said she couldn't remember what had happened, but as the years went by, and she told her story more and more times, she remembered more and more details, and in her old age – she died in 2010 at the age of 82 – she was irritated by people who doubted her story, as by now she could remember being taken by the eagle as well as trying to protect herself from further attacks in the hollow in the rock-face by throwing stones at the eagle. Strangely enough she never said she could remember the actual flight to the mountain.

Svanhild was weighed shortly after she was rescued, and the result of this has been a cause for several discussions over the years. In many books one can read that she weighed somewhere between 19 and 21 kilograms, but as Svanhild kept the torn dress for the whole of her life, this has later been shown to have been impossible. A 3½ year old girl weighing around 20 kilograms would have been seriously overweight – and Svanhild was tiny and slender all her life – and she couldn't possible have fitted in the dress. Probably the doctor misread the weight, as it was a weight normally used for weighing piglets and suchlike. Instead of 19-21 kilograms, Svanhild probably weighed 19-21 pounds (around 10 kilograms). Nevertheless, 10 kilograms is a serious weight for a sea-eagle to carry, and ornithologists uniformly agree that a sea-eagle would start to struggle with a weight of more than 3 or 4 kilograms, so 10 kilograms would be completely out of the question.

The great attraction
The doubts of various experts have in no way diminished the interest of the case today more than 80 years later. The local council have painted the rockledge where Svanhild was found to make sure that tourist can locate it easily, and in 1989 the local council adapted a new official coat of arms in the shape of a golden leg of an eagle on a red background. A new book on the case was published as late as 2006, but was roundly criticised the Norwegian ornithological society for misrepresentation of facts, downright fabrications and for scaremongering in relation to the sea-eagles.

What probably happened was that Svanhild just wandered off, and either dropped her shoe on the way, or threw it down from the ledge when she had walked out onto it. Although people at the time said it would have been impossible for a small girl to get to the ledge by her own, later tests have shown that it would as a matter of fact have been very easy. By the time the little girl was found, everybody was so convinced she had been taken by an eagle that they did not consider any other options. As for Svanhild's own testimony, it changed a lot over the years. She was probably told so many times by the people who had been out searching for her, that she had been taken by an eagle that she ended up believing it herself. Unless of course she wasn't taken by a sea-eagle, but by something even larger – this theory has in fact been put forward on several occasions, because according to legend, Norway is also home to what must be a local form of what is called the thunderbird in North America, a truly gigantic eagle so large, that it would dwarf even the biggest birds of prey known today. And it Norway, it is known as... the Gam!

A Norwegian rukh?
The gam is something along the lines of the Middle Eastern rukh known from the tales of Sindbad the Sailor, although not quite as big. The gam can't fly off with an elephant in each claw and one in its beak, it can only fly off with a fully grown bull,

but that is more than enough in my opinion. I know of only relatively few gam-sightings, and most of these are quite old and clothed in so much superstition and legend it is difficult to know what to make of them, but there are a few of a later date that show some interesting features. Most of these sightings are concentrated around Stetind, a mountain a bit further north in Norway. Stetind is a kind of national mountain of Norway. It is rather strange, looking almost like the drawings small children make of mountains. It is shaped like an almost perfect inverted cone, but with the tip cut off, so it has a very characteristic angular top, and it is up here the gam would nest according to legend. Today Stetind is a popular target for adventurous mountain-climbers, so there is no doubt about the fact that nothing nests on the top of the mountain. Nevertheless...

In 1967 two young men on a mountain-climbing expedition claimed they first saw a normal sea-eagle, and then a huge bird passing over the eagle, and making it look like it was the size of a jackdaw.

In 1976 a Norwegian woman claimed she saw a bird the size of a small plane flying past way up in the air.

In 1983 a Dutch tourist claimed a sighting of a bird so big it for a moment completely blocked out the sun.

There are very few details in these sightings, so I don't quite know what to make of them, but I do know that the clarity of the Arctic air makes it extremely difficult to judge the sizes of thing. Things can be considerably further away than you might think, so I am inclined to believe that what the mountain-climbers saw was in fact an ordinary sea-eagle with a jackdaw or perhaps a crow. If you are not especially interested in birds, those two corvids can look quite bird of prey-like (hell, they can even do so to a bird-watcher – I know I have been fooled many times, and I have been birdwatching for 40 years). The bird the size of a small plane was probably an ordinary sea-eagle again, but also a case of confusion size-wise, whereas the bird that blocked out the sun may simply have been a cloud.

The strange optical quality of the Arctic air sometimes have unfortunate consequences. I have on two occasions, once in Greenland and once in Norway, decided to walk to that mountain and that house over there, and much to my surprise have had to walk for a couple of hours before getting there. In each case I completely misjudged the distance.

The land of eternal light and darkness
With our visit to Stetind, we have to all extent and purposes passed into the Arctic. At Stetind we are north of the Arctic Circle, where for one day each year, the sun never sets, and for one day it never rises. The further north you go from there, the longer the periods of constant light and darkness becomes, until you reach the North Pole, where the only day of the year lasts for six months, and the only night for another six months. It is a strange world – I have never been in the Arctic in winter, but I have been there in summer, and I can vouch for the fact that it can be a very strange and unsettling experience.

Silver threads and glowing birds
The strangest and eeriest sightings I know of have all been made in deepest darkest winter. One was told to me by a friend who for some years worked at the University of Tromsø, one of the northernmost universities in the world. She claimed that sometimes in the middle of winter, when there had been a particularly magnificent show of the northern lights, you would see strange luminous threads or strands of something fall towards the earth. They would appear high up in the air, and then slowly drift downwards, all the while writhing and curling like earthworms on a wet pavement. But the threads never reached the ground. My friend, who is a fully qualified zoologist – she has a PhD in marine biology – would have been happy to sample the threads and analyse them, but despite numerous attempts, she was never able to catch one of them.

Locals also claimed that you would sometimes see luminous birds as well. Now there are not many birds in the high Arctic in winter, but some white birds, gulls, ducks, ptarmigans and suchlike continue to eke out an existence, and in deepest winter you could presumably see them fly by, glowing like subdued fireworks.

The people of the northern lights

And then of course, on very rare occasions, especially at times were the northern lights had been very powerful and seemingly reaching to the ground, you could end up meeting the light people – or actually see the light people – as far as anybody knows, no one has ever met them, but sometimes they have been seen far away, slowly walking across the mountains, or the ice, or sometimes even the sea. Always in small groups, looking for all the world like families dressed in faintly luminous fur-coats out for a walk or perhaps on a hunting trip. I have never been able to find any specific sightings of this light people, but apparently they have been seen a few times, perhaps once every 10 or 20 years.

I must confess I have no idea what these various luminescent phenomena can be. I can only speculate as to whether they are some kind of optical, perhaps electrical phenomenon triggered by the northern lights, as these things apparently only are seen in connection with that particular natural marvel, but your guess is a good as mine.

The northern lights are a strange and magical phenomenon, and although they are fairly well understood, they are also in a strange way otherworldly. If you have ever seen northern lights in its fullest glory, with swirling coloured curtains rushing back and forth across the sky, you will never forget it. And if you grow up with it, I have no doubt it will be an integrated part of you, because there are in a sense a real people of the northern lights. They are the Sami. They are a people all to themselves. They have their own beliefs and mythology, and sightings of beings that are completely their own. The Sami live on Norway, Sweden, Finland and northern Russia, but to avoid treating them in three different chapters of this book, I will look at all their stories in the chapter about Sweden.

SWEDEN

SWEDEN - So near and yet so strange

For most people the name Sweden conjures up a vision of endless dark and broody conifer forest, and gloomy people spending most of their life making up ridiculous rules and regulations. To some extent this is not very far from the truth, and it can sound dull and uninspiring, but in most cases Sweden is a land of great natural beauty, although the conifer forests tend to get a bit repetitive when you go for a drive or on a train journey through the countryside.

For me growing up in Copenhagen with a father who worked in the Swedish town of Malmö, Sweden was an almost magical country. Because of some weird differences in taxation, stuff like chocolate and marzipan was much cheaper when bought in Sweden, so my father always brought goodies back home, especially around Christmas, where marzipan was brought in blocks of one pound each. Actually one of my first cryptozoological memories revolves around the time – I think I was 7 or 8 years old – when I made a plesiosaur entirely out of marzipan on Christmas Eve, and then ate it all in one go. I did get a severe stomach-ache, but I still love marzipan.

When you go for a drive around the Swedish countryside, things can seem a bit lifeless at first, but on closer inspection there are lots to see. Sweden has a rich animal life with lots of big mammals, moose, bear and wolverine to name but three. The birdlife is rich and varied as well, with large birds of prey like golden eagles and white-tailed sea-eagles, and massive owls like the great horned, and the great grey.

But of course there are other things lurking in the shrubbery. The Swedish forests are big and dark, and there is plenty of room for strange creatures. Up north there are endless wide open spaces where you can see a bear a mile off, but nevertheless, critters manage to hide themselves up there as well, and even in the extreme south, which is intensely cultivated and densely populated, there are lots of things going on.

The only problem is, it can be a bit difficult to get the Swedes to talk, especially when a Dane comes asking stupid questions. It helps if you speak the language of course, but you do sometime hit a very solid wall of suspicion when you start asking people whether they have seen any strange animals. It is annoying but perhaps understandable as I once met a lady who had been fired from her job as a crisis counsellor because she told her boss about a black panther she had seen in the forest outside the town of Gothenburg. Her boss was of the opinion that if she claimed she had seen a big cat, she had to be mentally unbalanced, and as such not suited to help people with problems. And I once got kicked out of a youth hostel when I tried to get the couple who ran the place to talk about trolls! The perils of being a researcher of weird things!

Sweden has a rich and varied folklore – to some extent nothing but a variation on Danish and Norwegian stories, but also decidedly different, not the least because the northern parts of Sweden – Lapland, or Sápmi, as it is called today (which also includes parts of northern Norway, Finland and western Russia) – is home to a people with a language, culture, history and belief system quite different from the typical Scandinavians living further south. The Sami people tell stories about all kinds of weird creatures, spirits living in rocks and trees, and a veritable cornucopia of gods – not all of them benevolent. So the stories about strange animals and possible otherworldly beings vary quite a lot as you go from one end of the country to the other. I realize that Sápmi actually covers 4 different countries as such, but to avoid repeating myself *ad nauseam*, I will treat all of it here.

Sápmi
The Sami people have always been seen as strange, sometimes heathen and even slightly scary. When the Swedish naturalist Carl von Linné, who invented the taxonomical nomenclature we still use today in zoology and botany, went for an expedition to Lapland in the 18th century, he described it in much the same way as if he was off to the South American jungle to visit headhunters or something equally exotic.

Unfortunately not a lot has been written about the Sami people, not the least for linguistic reasons, but you can find the odd story here and there, and every so often you meet people who are more than willing to tell you stories till your ears starts to steam. In my case it was Mari and Gunnar Åkesson whom I met in the little town of Arvidsjaur in Northern Sweden during one of my trips, when we were staying at a youth hostel called Lappuglan (The Great Grey Owl). Mari was Sami and Gunnar was Swedish, and they were happy to find an eager listener (I filled two complete notebooks during my stay!). The following is an extract of our many hours of talking. I have inserted eye-witness sightings when I have been able to find them.

Golden, flying, burning and steaming reindeer

Reindeer are without comparison the most important animal in Sami life. They have been herding reindeer for thousands of years, eaten them, used them to pull sleds, used their skins, drunk their milk and what have you, so it is no wonder that there is an enormous amount of stories and legends about reindeer – and a quite impressive number or words about reindeer, their behaviour and their relationship to snow in the Sami language. Normal wild or semi-domesticated reindeer are fairly small, rather dopey deer, with a set of antlers so large they would comfortably fit on a full-grown moose – and the females have antlers too! But, there are also some rather strange reindeer out and about, and every now and then, you may run into one of them.

Meandas-pyyrre is the most legendary of the lot. This huge animal has a direct link to the many gods of the Sami, so one would not expect to meet it in person, which is a bit of a shame because it sounds like it would be quite a sight. It is extremely large (naturally), with golden antlers, a black head, white body and burning eyes. To some extent is sounds very much like similar legendary godlike beasts from others areas of Scandinavia – Thor has a massive pair of goats to pull his chariot, and they have burning eyes. The troll-hunting kings of Denmark rode horses with burning eyes and had hunting dogs with the same ocular equipment. Burning eyes is apparently the in-thing with slightly demonic animals.

So what about the more flesh-and-blood-like reindeer one bumps into on a regular basis in the land of the Sami? Most of them are of course straight forward standard reindeer, but not all of them.

In the summer of 1990 I heard about several sightings of a rather strange looking reindeer in the vicinity of Gällivare, a town in the far north of Sweden. It is a very popular resort for cross-country skiing in winter, which basically means, that in summer there are lots and lots of walking pathways to use to your heart's content. It was in this area that at least six different people claimed to have seen a white, or rather cream-coloured reindeer with yellow antlers and not exactly burning eyes, but red eyes. The animal was extremely wary, and the people who had seen the thing, had only got a fleeting glimpse of it. Nevertheless, the last sighting had occurred only two days before I arrived in Gällivare, so I decided to try and see the animal myself. But although I spent a lot of time out and about, and sacrificed a substantial percentage of my blood-volume to the more than eager servants of the local mosquito god, I had no

luck, But I managed to track down and talk to one of the eyewitnesses to get an idea of what he and others had seen. It was at the same time very strange and extremely ordinary – just a reindeer trotting across an opening between the trees and disappearing – but said reindeer had a very strange colour as described above. I can only suggest that it was a reindeer with some form of genetic disturbance in the colour department. Leucistic animals often have a very strange washed-out look, and I think this is what people had seen. I have at one time seen a wren with this same genetic aberration, and it had the same colour as a cup of tea with a bit too much milk in it.

A few years later I got a letter from a friend who had been walking through Sarek Nationalpark about 60 km west of Gällivare. He said he had met some locals who claimed the area was home to a very special breed of dwarf reindeer. Now I know there can be quite a difference in size depending on where you encounter your reindeer – the Greenland ones are fairly big, whereas the most southerly living animals are distinctly on the small side. But these were supposedly only the size of sheep, but with fully developed antlers. It sounded very unlikely, but when I started digging, I found plenty of people who claimed to have heard about these dwarf reindeer, but not a single one who had seen the animals. I still haven't, so I am inclined to think that they are some kind of urban legend as it were, although urban is not a word that comes easy to mind when you think about the Sarek area. It is a very beautiful and very remote area, and not a place to go walking unless you are well and truly prepared. You may even run into a bear, and they should be treated with the outmost respect.

For the final kind of reindeer we are almost certainly back in the land of legend again, but nevertheless. The Sami talk about the Juovlajohttit, the Christmas travellers, which are spirits on sleds that only some people can see. The move across the sky, and according to some of the stories, their sleds are sometimes pulled by flying reindeer. In a few cases they are even dressed in red, and this, after the Coca Cola company got hold of the story, should in fact be the reason why Santa Claus is dressed in red and flies around in a reindeer-powered sled at Christmas. I am inclined to think, that this story is based on a form of mirages caused by the northern lights, but I have met an old man i Luleå at the Swedish East coast who were adamant that he had seen the travellers when he was a boy.

But perhaps he was just one of those people, who could see the otherworldly beings. He also claimed to be able to see the kadniak, a special kind of little people. They are small, dress in red and have very long hair. They are closely related to the gufikktar, a kind of fairies or elves who are by and large very nice, but – and the old man looked me straight in the eye – should I met one of them and be invited to a party, I should under no circumstances eat or drink any of the things I would be offered, because if I did, I would be caught in a form of time-warp, meaning when the party was over, I would think a few hours had passed, but outside of the great hall where the party was being held, many years would have passed, and all my loved ones would be dead and long gone.

Since I am sitting here writing this, I think I can safely assume that the people I met subsequently who asked to buy me a drink, were in fact quite ordinary people. Pheww!

Danger, danger, danger!
When you are out and about in the wilderness in northern Scandinavia, you have to know what you are doing, and you have to be able to look after yourself. There are things out there that can be dangerous. The weather can kill you quite easily, and there are animals you should treat with the outmost respect, and there are creatures you should stay as far away from, as you possibly can.

I have already covered the various forms of trolls to be found in Scandinavia, but among the Sami people, stories are being told of other humanoid creatures that are quite different, and far more dangerous. And even today, people claim to see these creatures on rare occasions.

Big, lazy, stupid and dangerous
Quite a lot of stories concern the stalos, giants of a rather disreputable disposition. They are not only far too big for comfort, but they are also man-eaters. Luckily they are also stupid and lazy, so even if you should stumble into a couple of them, you are not necessarily in mortal danger. But you should of course have your wits about you, especially if you are forced to camp out in the wild. Never sleep too soundly, or you may be crunched before you know it.

These stories are interesting, but clearly folkloric, as there are lot of variations over the theme of a small and intelligent boy fooling the big and stupid stalo, so I will let them stay where they are, and concentrate instead on the snowmen – and they are even more interesting.

The snowmen – and forgive me for thinking yeti, bigfoot and suchlike at the moment – are big hairy apelike, but also vaguely human-like beings covered with a very thick, but also extremely malodorous fur. They have very large feet (somehow this sounds familiar as well), and they live in caves. They love berries, especially cranberries, but they can also kill the odd reindeer and eat that.

This to me sounds very much like a living thing, might even call it a brown bear. The stories about them are also refreshingly free of any kind of magical element, so what can they be? It is also quite interesting that in a few cases people have found very large excrements filled with berries, something that would have made me think bear straight away (again), but according to some of the eyewitnesses, the droppings are very different from bear droppings.

A couple of sightings to demonstrate what I am on about would be in order round about now I think. I should however say that very few, if indeed any of these sightings can be found in print. According to Mira and Gunnar mentioned above, most of the people they know who have seen strange things, would never ever talk about it to outsiders, and definitely not to newspaper reporters, so for the sake of completeness, I will bring all the sightings I have been able to collect (only 6!) in full, as they, to my knowledge, never have been published before. There may be a few others out there, but I have never been able to find them.

All of the sightings come from the area around Kebnekaise, the highest mountain in Sweden. It is located in a mountainous area west of Kiruna and north of Stora Sjöfallet National Park. According to the people I have talked to, the sightings go back several hundred years, maybe even thousands of years. It is possible they were in fact making a fool of me, but several of them claimed that the Sami people were fully aware of the existence of these creatures, but saw absolutely no reason to involve anybody else in them. They wouldn't understand anyway.

This concept of "wouldn't understand" was something I heard on a regular basis. Many Sami people claim not only to know something about the nature and wildlife that surround them, but to understand it on a far more fundamental level than outsiders, an understanding that can only come from living in close contact with the nature of the area for centuries, generation after generation.

Sightings

1953. Early spring. A Swedish mountaineer on his way towards the top of the mountain(Kebnekise) saw what he described as a family of white gorillas. In a letter to a friend [which I have seen and read in 1996 – it is now in the possession of the grandson of the original eyewitness] *he wrote* [translated from Swedish]: *"I was going up the easy route, when I suddenly saw something moving in the distance. As I got closer I saw it was a group of large white monkeys that looked like gorillas. It all looked so natural that it took me several seconds to realize that this was not normal. There were no gorillas on Kebnekaise! So what were they? I had no idea, and when I tried to get closer, they saw me and disappeared. Maybe we have our own snowman in Sweden? What do you think?"*

1961. Middle of summer. A Sami student from Stockholm was walking towards Kebnekaise from Nikaluokta about 10 km to the east, when he saw a large white figure running very quickly across the bottom of the valley about 3 or 400 metres in front of him. He only saw it briefly, but said that it looked very humanlike, walked or ran upright on two legs, and had shaggy white fur, almost like an "unwashed polarbear".He described his sighting to several other people he met during the next few days, but none of them had seen anything like the creature he had seen.

1969. 12th September. Two German tourists staying at the Kebnekaise Mountain Lodge on the south side of the mountain claimed they saw two polar bears off in the distance towards the mountain. When questioned closely later on, the told that they had seen two very large white animals in the distance. They were crouched down, and seemed to be looking at the ground or perhaps resting. The only thing they could think of being that big and white, were polar bears. [Although polar bears are indeed the quintessential Arctic animals, they are not found in Sweden, and they are definitely not found around the Kebnekaise mountain range].

1971. Early June. A sami woman and her two young daughters were driving westwards towards Nikaluokta on the road along the northside of the Laukkujärvi Lake. When they were about 20 km east of Kebnekaise, the woman slowed down the car because her youngest daughter complained of being carsick. "I was trying to comfort my daughter, and trying to determine whether she was going to be sick in the car, when I suddenly had to step on the brakes and just sit and stare. Suddenly this huge animal was standing next to the road looking directly at my car, as if it was trying to dare me to drive on. My girls started getting scared, and the youngest started to cry. For a moment I thought the animal would attack us, but it just grimaced, almost like a smile, and then it walked across the road and disappeared towards the lake. It was only 10 metres away, and I could see it clearly. It was bigger than the biggest man I have ever seen, it was very white, it had big yellow eyes, and fur that looked like it hadn't been washed for years. It smelled very bad too. Even though the windows in the car was closed, we could clearly smell it. It smelled like a dead reindeer lying in the summer sun."

1984. 18th July. Two Danish birdwatchers on a visit to Abisko National Park about 50 km north of Kebnekaise had been walking around the Abiskojaure Lake for several days, and had seen a brown bear the day before. Sometime during the afternoon they came to an area with a lot of low but very dense bushes. Around noon they stopped to have lunch as best they could for the many mosquitoes. While they were eating, they suddenly heard a deep growling sound close by, and thinking they had somehow stumbled onto a bear, they quickly got up, and started to walk slowly away from the area. In their haste to get away they forgot the remnants of their lunch, and when they looked back, they saw "A bloody great bear eating our sandwiches! Only thing was, it had the same colour as double cream, and it seemed to have very long back legs. I almost thought it looked like a hairy human, but my friend told me not to be ridiculous. He absolutely forbade me to go back and have a closer look, so we left, only remembering our cameras half an hour later. By then it was too late. I still think about it every now and then. I have never heard of an abominable snowman in Sweden, but that's what I think it is."

1991. 4ᵗʰ August. *4 English tourists on their way to the top of Kebnekaise stopped to admire the view. After a few minutes one of them realized that a bit further up the path, another figure was looking down towards them. This figure was rather strange looking. First of all, it was clearly a very big man, definitely more than 2 metres in height, and broad as well. One of the tourists suggested it was in fact actor Richard Kiel, the man who played Jaws in several James Bond movies, who were out walking as well, and that they should all get his autograph. Unfortunately they never did get a chance to get any autographs, because suddenly "Richard Kiel" started running down the slope with lightning speed, and very quickly got lost in the distance. None of the tourists thought about taking a picture although they all had cameras, but afterwards they described the figure as very humanlike, but with a short whitish fur.*

This forgetting-to-take-pictures-although-having-camera syndrome is something you hear again and again when it comes to strange phenomena, fra cryptozoology to ufo's to ghosts to just about anything weird. Many authors have tried to read something slightly sinister into this, as if the phenomena somehow could adversely affect cameras and/or the witnesses themselves. Personally I see no reason to invoke strange forces like that. Surprise and fear is more than enough to make people freeze, forget their equipment or simply become clumsy and butterfingered if they get as far as trying to take a picture. It happens to people who take pictures of birds and other animals, so why not to people experiencing things even more strange?

Yikes! It's the padnakjunne!!
As if things aren't bad enough with the snowmen, things get absolutely gothic when we get to the padnakjunne. That is a tribe of cannibalistic humanoids with the face or snout of a dog. Nasty, viscious and extremely dangerous beasts. I don't know much about them to be honest – I couldn't find anybody willing to tell me anything about them, let alone claim a sighting or something, but thanks to a Sami friend of a Norwegian zoologist I know through a Danish colleague (yes I know, it almost sounds like a friend of a friend story, but I have it in writing), I have managed to get a little information about them.

The padnakjunne comes from a time where wolves were considerably more common than they are today. Compared to say 2- or 300 years ago, the wolf population of contemporary Norway and Sweden is absolutely minute. But back in the day, as it

were, wolves were a very real problem if you were farming animals in any way, be it cows, sheep, horses or indeed reindeer. And of course there was the fear of attacks on human which despite what especially many modern people seem to think, never has been a big issue. Domestic dogs, horses, bulls and suchlike have always killed and maimed far more humans than wolves, but that hasn't stopped modern Danes, Swedes or Norwegians to completely lose it and demand the eradication of any and all wolves should they be seen even remotely close to what could be called civilisation. Some years ago when a young wolf went walkabout from the small wolf population that lives in southern Norway and western Middle Sweden, and wandered southwards through Sweden, people absolutely panicked, and started clamouring for school-buses with armed guards, even in the middle of major cities, and there were stories of people in the same cities being afraid to go out after dark because they were expecting to be attacked by the wolves. *"They are cunning you know! And bloodthirsty!"* It is hard to raise a voice of reason in cases like this, and I have been threatened with physical violence more than once in these cases.

Where there are wolves, there are also, or so it seems, werewolves, or at least some kind of strange in-between beast, half wolf and half man, sometimes simultaneously, and sometimes mostly man, and only wolf at particular times. The Sami werewolf, the padnakjunne, was precisely one of those creatures, although in their cases, there was no mentioning the full moon or anything like that. They were a special tribe of beings with a penchant for human flesh. They were only interested in hunting humans, not in any other kinds of animals.

Today the padnakjunne are long gone. According to my Sami correspondent no one had claimed a sighting of them for at least a couple of hundred years , and he was inclined to think that they were in fact some sort of bogeymen invented by the adults to scare children from running around at night or straying too far from the houses or the camp. He even remembered his grandmother threatening him when he was young (and a bit unruly) that if he didn't behave, the padnakjunne would come and eat him.

"I never did learn to behave, but they never came!"

Weatherbirds
I have already discussed the very large eagles of Norway, some of which have been seen a few times, or so it is claimed, in Sweden, but there are also stories about another kind of big bird in the northern lands, which is worth a closer look. I don't know if they have a special name – if so I have never been able to find it. And it is also hard to refer them to any special group of birds, as the stories and descriptions I have heard of them are frustratingly vague when it comes to detail. But basically it is creatures that very much brings the North American thunderbirds to mind. It is extremely large birds – very large birds – that you only see once or twice in a lifetime, or perhaps never. But on rare occasions when a lone wanderer happens to

glance upwards, he or she may see an enormous birds gliding past with hardly a wingbeat. It is always so high up you cannot make out any details, but it is something big enough to block out the Sun. And it is always followed by weather of the very worst kind imaginable – snow, rain, cold, storms, you name it, and the heavens will lash you with it.

All the stories I have heard about these bad weather-birds, have at least been third or fourth hand, so I am in no position to judge their reliability in any way, but I tend to think that what people have actually seen are the first dark clouds of an incoming storm or similar types of bad weather. Perhaps these clouds occasionally have taken on the shape of a very large gliding bird, and in a time and amongst a people who had no idea how weather was formed, and knew nothing about air-pressure and cold fronts, it was easier to blame bad weather on some very big and very scary birds, or simply see them as portents, omens of things to come.

With this little story of the weatherbirds, it is time to leave the Sami territory behind and travel southwards through the mighty conifer past Stockholm and all the way to the south coast of Sweden, where the climate and the landscape is considerably more gentle, but where weird things still happens. We also have to call in on a petrified giant, and try to find out what happened to the headless seals at Måkläppan.

The mighty owl
In northern Sweden the forests are thin, and in some areas completely missing, but as you go further south, the trees start to crowd in on you, and before you know it, the real forests starts. They are big, dark and scary, but we will come to them in a little while. Before we get as far as that, we must first try to avoid an immense owl, slagugglan – the strike owl. And that's a bird you want to treat with the outmost respect and caution.

Now Sweden is already home to a handful of big and rather awe-inspiring owls, the eagle owl, the great grey owl and the ural owl. The bird that should interest us here is the ural owl. It looks basically like an overgrown tawny owl, and it is strangely enough also called slaguggla – the strike owl in Swedish, because even though it is in fact only a medium-sized bird (50-60 cm in length), it has nasty temper, and a reputation for ferocius attacks on anything or anybody than comes close. Ornithologists wanting to study and band ural owls have to work wearing motorcycle helmets, thick leatherjackets with the collar turned up, so as not to get their necks lacerated, and heavy gloves to protect them from the owl extremely sharp claws and beaks.

The "real" strike owl on the other hand is quite a different kettle of fish (bowl of owl?). According to the various stories about it, it is the size of a grown man, has a wingspan of about 3 metres, and is capable of killing and flying off with a reindeer calf. Now there's an owlman if ever there was one!

I am in fact using the word "owlman" very deliberately, because there is a striking similarity between some of the sightings of the strike owl and the notorious Owlman allegedly seen around the small church at Mawnan in Cornwall in the mid 1970's. The strike owl has been taken to be a man on several occasions, or people have told stories about seeing a man with glowing eyes sitting in a treetop before taking of and disappearing on big broad wings. Some people claim to have been attacked by the big owl, or at least having it flying directly towards them, and only veering off in the last possible moment.

Most of the sightings that I have heard of in relation to the strike owl are from the northern half of Sweden, i.e. north of Stockholm, and they correlate quite precisely with the distribution of especially the ural owl. How the stories of the owl flying off with reindeer calves have started, I have no idea. I have never been able to track them back to an original source or sighting. It seens to be "just so". Another thing that strikes me with the sightings and stories of the strike owl that I have have collected, is the fact that most of them (17 out of 21) are made by children or teenagers. This fact

may seem of no consequence, but I think it of the outmost importance regarding the origin of the strike owl. If you are say 10 or 12 years old, perhaps out on a day with bad or overcast weather, and are suddenly attacked by a 50 cm owl with a wingspan of more than twice that, you are extremely likely, I would say with the outmost certainty, to misjudge the size of the thing quite badly. Having myself experienced the ferocity of an owl attack myself, although in my case only a long-eared owl, I can vouch for the fact, that you feel you are being attacked by a monster at least as big as yourself. I plain and simply think that the strike owl, the Swedish owlman if you like, is nothing more than the actual strike owl, the ural owl, dressed up and enlarged by fear and shock, and maybe the odd exaggeration, when the story of the attack had to be told for the umpteenth time.

A different set of wings

While we are on the subject of winged beings, there are also a couple of histories of winged cats from Sweden – and one from Denmark as well, so I might as well deal with them here. Winged cats are a subject that has been studied and described very thoroughly by British cryptozoologist Karl Shuker on his website and in several articles in magazines like *Fortean Times*, and very thoroughly in a complete chapter in his 2008 book *Dr. Shuker's Casebook.* Winged cats are plain and simply cats – usually normal domestic cats that seem to have wings. The stories about them are sometimes rather dramatic, some claiming that the cats although not entirely capable of flying, at least thanks to the wings can perform extremely long and high jumps. They are also in some cases described as extremely aggressive, capable of attacking without any kind of provocation. The same goes for one of the Swedish winged cats – it was actually shot and killed somewhere in central Sweden because it attacked a child. The exact place, date and year is unknown as far as I have been able to ascertain, but it was either in 1941 or in 1944 during the Second World War, which might explain where there is relatively little information to be found about it.

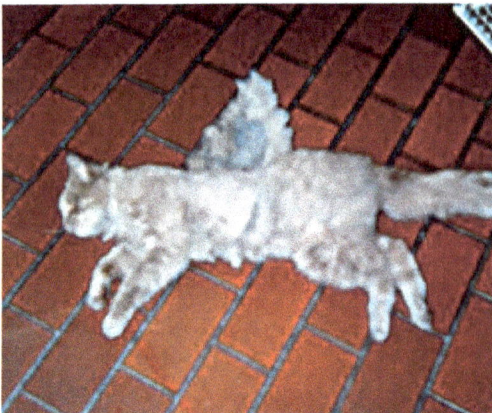

In all the cases were winged cats have been thoroughly examined, they have of never shown any sign of having real wings. The "wings" were either long mats of fur, or caused by a rare condition that makes the skin of the animals extremely stretchy and capable of being drawn out in long lobes. In some cases people claimed to have seen the wings actually flapping, but that has probably been nothing more than an optical illusion.

The winged cat that was seen and later caught in the Danish town of Frederikshavn in 2010, was a stray. Several people had seen it, and when it was jumping, it seemed to have a couple of loosely flapping wings on its back, although to some it looked liked the cat had entangled itself in a scarf of a piece of fabric. The cat was finally caught by an animal welfare inspector and the wings turned out to be only one wing consisting entirely of matted fur. The cat also turned out to be in an extremely poor condition, so it was put down, but the wing was unfortunately not kept.

The second case of winged cat is from roughly the same period of time. From 2009 to 2012, a fairly large greyish cat with a couple of loosely flapping wings were seen on several occasions around the little town of Börringe in southern Sweden. Nobody claimed ownership of the cat, and since Börringe is located in the middle of an excellent area for birdwatching, I kept an eye out for it everytime I was in the area, but I never managed to see it. I interviewed several local people about it, and they all claimed that the wings were a seasonal phenomenon, or perhaps not strictly seasonal, but that the wings of the cat would get gradually longer over a period of time, and then suddenly one day the cat would be seen without wings, and then the wings would start to form again. I think this clearly demonstrates that at least the wings of this particular animal was made of fur, and when they reached a certain length, they would either simply drop off, or get caught in vegetation or similar obstacles and be ripped off. And since this phenomenon would repeat itself, it seems likely, that the cat may have had a particularly tangled coat, perhaps due to some form of genetic aberration, which would facilitate the formation of the wings.

The problem with big cats

Strange though they are, winged cats are rarely especially big. But big cats as such, or rather sightings of big cats in areas where they definitely shouldn't be, are in a sense the mainstay of modern cryptozoology. There are probably more sightings of Alien Big Cats, or ABS's, than all the other cryptids put together. Several books have been written about this subject alone, but strangely enough, apart from Denmark, this has never been a big thing in Scandinavia or in the Baltic Countries. This is probably due to the fact that the winters in Denmark are fairly mild compared to the other countries, and thus making it just that little bit easier to survive for a big or medium-sized cat on the prowl. An additional problem is the fact that most of these countries already are home to, if not a big cat, so at least a medium-sized cat, the common lynx. Even though a lynx is a relatively characteristic animal with its long legs, tufted ears and very short tail, it can be extremely difficult to sift through cat sightings and decide how many of them are something interesting and how many are just lynx sightings.

Lynxes are notoriously shy and secretive, and you can in fact live in an area with a lynx population for years without ever seeing the slightest trace of them apart from the odd footprint every now and then, but you can also see them on a regular basis. I personally think it has something to do with the individual animals, some of them shy, some of them media whores so to speak. But every now and then, you do get sightings of cats that are clearly not lynxes, i.e. people very clearly describe their long tails. So what are they?

I have no doubt that most, if not all sightings of ABS's are in fact sightings of escapees, animals that have been held in captivity, perhaps as some kind of exotic pet – sometimes legally, but sometimes absolutely not, which naturally means that the owners are reluctant to report any break-outs. Keeping big cats in captivity without a whole bunch of permissions will give you substantial fines, and can in extreme cases land you in jail.

To give a general idea of the phenomenon here is a bunch of sightings from Sweden:

- A fairly small but heavily patterned cat looking like an ocelot or perhaps a margay was seen twice in the vicinity of Falkenberg in the summer of 1981, and once crossing a road in the centre of Laholm at the end of September same year.

- A golden-coloured or dusty yellow cat looking like a small lion, but the size of an alsatian was seen running across a field near Tranås in 1974.

- A very large black cat was seen late at night running along the main coastal road near Karlshamn on the south coast of Sweden on May 5th 2009.

In 1999 I received a letter from a family in Hudiksvall who told me that sometime in the beginning of the 1960's – they couldn't remember the exact time, but it was in late spring – three members of the family had on three separate occasions seen a leopard in the vicinity of their house. On all three occasions the animal had been sunning itself on a big rock close to their house.

In 1985 around Eksjö and Nässjö in central South Sweden, there were a number (8 that I know off) of sightings of a large brownish cat, with a long tail, short and powerful legs, and a fairly small head, presumably a puma. All the sightings were within two weeks of each other, and then the animal disappeared as suddenly as it appeared. It was probably shot or recaptured by whoever lost it in the first place.

In a letter I received at the end of 2013, a boy living in Götlande told me he had seen what he described as a very big grey and black cat with a very thick and shaggy fur, and a thick bushy tail. He thought it was a least as big as a lynx which he had seen in Kolmården Zoo outside of Stockholm. He was also very interested in cryptozoology, and was certain he had seen something very much out of the ordinary. Personally I think he had seen a free-range Maine Coon cat or something similar. There are a couple of domestic breeds that can get extremely large, and typically have a very thick shaggy fur. Wild cat species are usually considerably more sleek-looking.

And finally a girl living in Karlstad told me how she had seen a clouded leopard in a small forest close to her home in the summer of 2001. She described it in precise details, so I have no doubt she actually saw what she said. The animal was climbing an old tree, possibly hunting for birds or bird's nest, and the girl said that it at one stage was climbing almost upside-down along a very thick branch, something clouded leopards are in fact fully capable of. They are probably the most arboreal species of cat.

The weird ones
I have no idea how many cryptozoological books I have read over the years, but it is a fair few. In almost every one of them there has been a section of sightings or creatures that doesn't really fit into any of whatever categories the particular author has been using, but the stories have still been too good to not include them in the books – and why should this book be any different? So here are a handful of strange stories that I like. I have no idea why I am presenting them at this place in the chapter – probably because I couldn't think of anywhere better to put them. So here goes...

Death in the summer
The first of the weird stories are in a sense not even cryptozoological, but it is still so mysterious that I include it here, if nothing else for the opportunity to send a little chill down the spine of any reader. So off we go to the extreme southwest of Sweden, to Måkläppan, a natural reserve off the coast of Falsterbo, one the best places to watch migrating birds in the autumn in northern Europe. Måkläppan is excellent for ducks and waders, and it is also rather popular among the local population of seals, which can sometimes be so big the birds can have trouble finding a bit of free sand to stand on. I have spent many an August or September afternoon staring out at Måkläppan through my binoculars, counting birds and seals, but have in fact never seen anything more strange than a few rare birds every now and then.

In the summer of 1984 a considerable number of the seals on Måkläppan died, possible because of some kind of disease epidemic. It is something which has happened several times before that – and since, and as such there is nothing mysterious about that. It has happened on several occasions in Denmark as well, and in some instances has devastated the seal populations. But within a decade or two, the number of seals iae usually back to what it used to be.

The mystery of 1984 became clear, when the authorities decided to collect a number of the dead animals to make tests and decide what kind of disease had killed them. All well and good except for the rather eerie fact that a large number of the animals were headless.

This may sound extremely unmysterious, as all dead animals will start to fall apart sooner or later, and that is indeed so, but seals are quite sturdy in construction. The heads doesn't just fall of, and the skulls do not disappear like that. They may be small, but they are thick and heavy and not something a bird like say a crow or a gull would fly off with. So what the d.... happened? I have absolutely no idea, and as far as I know that goes for everybody else in these parts. But I am as always open to suggestions if anybody has any ideas. There has of course been some conspiratorical speculations, but with absolutely no evidence – government experiment with biological warfare, mutilations done by aliens – just like the cattle mutilations in USA, and of course satanic cults!

The black ones
Black dogs have always been a mainstay of cryptozoological literature, although I am not certain of whether the dogs are in fact true animals in a physical sense or whether they are some form of ghosts of zooform phenomena as named by Jonathan Downes. But alas I am not here to discuss the physical reality of Swedish black dogs, as I have never been able to find a story of a Swedish black dog. There might be some, but for some reason they have never come my way. But one thing that Sweden does have, and which I am quite sure is absolutely unique, is a black moose, or at least a handful of stories about a very large, very silent and very dark moose haunting the roads north of Kristinehamn and Karlskoga west of Stockholm. All together I have seven sightings, covering 8 years from 1994 and onwards, of what the eyewitnesses have described as an enormous moose, that suddenly appears in from of people, just stands there looking at them, or rather often looking down at them from a great height, and then after a few minutes starts to drift away through the forest or along the road without making any sound what so ever. At first the experience is extremely scary, but within a few seconds a strange calms descended on the people, and they felt happy and contented, as if the moose somehow communicated the fact that it was keeping an eye on them, and would make sure they wouldn't come to any harm. One of the eyewitnesses I have talked to even described is as en experience that changed her life forever.

Although I have the greatest respect for life-changing experiences, and am in no way trying to diminish what the eyewitnesses have experienced, I am quite convinced that what they saw was in fact nothing more and nothing less than a perfectly normal moose. If you are not used to them, meeting a moose in its natural habitat can be quite a shattering experience, and it is interesting that all "my" eyewitnesses" were tourists with absolutely no experience of moose in the wild or for that matter in zoos and similar places. A moose is a shockingly big animal, especially a male. Even without its antlers it will still look down upon you from a height similar to a big Shire-horse or something similarly enormous. But the most extraordinary things about a moose is its ability to move through a forest filled with dry twigs and wilted leaves, and not make more noise than a snowflake. I have seen moose perform that little trick on several occasions, and it is completely magical. Your mind simply refuses to accept that an animal this big can move so silently. And then of course there is the personality of the animal. Moose are not especially frightened of humans and their cars, which far too often leads to a rather messy end, and they look upon you with an expressions of gentle gloom that makes it very difficult not to try and comfort them, and telling them everything is going to be all right. They seem incredibly gentle and sad. I am of course anthropomorphising wildly, but that impression can be extremely vivid and give the animals and almost otherworldly feel. On the other hand it is a bit strange, that considering the fact that there are moose in most parts of Sweden, and that a lot of tourists visit the country, I haven't got more sightings like that. Maybe it is just so, and I can't really offer any explanation except perhaps that the area of the sightings for some years were the home of an especially friendly moose with an especially soulful expression.

Beware!

Perhaps I should after all end this section with a little warning. Although moose are very charming and by and large fairly friendly animals, they are not to be trifled with. They are very big, and if they so wish, they can be quite dangerous, and should be treated with care and caution, especially if you are a driver. Thanks to their long spindly legs, things will very quickly turn nasty if you crash into one of them. The car will just swipe the legs out from under the moose, and the body, which in big males can weigh half a ton, crashes through the windscreen and ends up in your lap. Not a good way to end a Sunday drive!

Try me kangaroo etc. etc. etc

There is of course no lack of stories of escapees in Sweden. Just like in almost any other country you can hear stories of escaped animals from zoos, from private parks, from farms and what have you. Most of these are fairly straight forward, and most of the animals are caught, or in some cases killed, fairly quickly. But one thing baffles me – why do the majority of these sightings all concerns kangaroos? Is Sweden home to some sort of secret kangaroo-worshipping society or is the keeping of kangaroos and wallabys as pets fare more common that I would have thought?

In cold blood

Like the other countries in Scandinavia, indeed like most other countries in these northern latitudes, the number of reptile and amphibian species in Sweden is fairly small. A couple of lizards, a couple of snakes, a handful of frogs and a few newt and salamanders, and all of them, at least officially, rather small. And yet people insists that they on quite a regular basis see members of both these animal groups that definitely shouldn't be roaming the Swedish countryside.

Sightings of exotic reptiles is a bit difficult to handle in this day and age where just about everybody seems to be keeping pet pythons, domestic geckos or tame turtles, so when it comes to modern sightings, I tend to think that most, if not all of these sightings are of escaped or deliberately released pets, in most cases probably animals that have become too large to handle, or maybe the owner have simply lost interest and can't be bothered to make sure the animal finds a new home. But some of the sightings are so strange they are definitely worth a closer look.

Lizardmania

I am perfectly willing to entertain the idea that one or two Swedish people over the years could be clumsy or even stupid enough to lose their pet lizard, but at least seven of them? All over the country? And all during the same summer? No way! Nevertheless, that apparently was exactly what had happened in the summer of 1993, when sightings of large lizard (1 metre in length or even at bit more) started to come in from at least 9 different places in Sweden. And not only that, the animals were highly aggressive if you ventured close to them. According to the descriptions they looked like your typical large monitor lizard type animal, like the African Nile monitor or some of the big Australian goannas. They are not your run of the mill pet, not the least because they do grow to be big and usually with a rather nasty temper and a set of rather formidable teeth capable of doing quite a bit of damage.

As far as I know, there is no "official" statistics of how many exotic pets escape, probably because a lot of the people that actually loose animals like that, would be rather reluctant to report in anywhere, but I am having a hard time believing that so many animals actually escape. Keeping exotic pets are not something for the green beginner, so one would expect most of these people to actually know what they are doing. I once wrote an article about exotic pets to a Danish weekly magazine, and during the writing process, I interviewed the president of a society of exotic pet keepers, and according to him (and I don't think he was trying to hide anything from me), escapes are in fact rather rare, and when they do happen, it is mostly snakes that do a runner. The only escaped lizard he had ever heard of was a blue-tongued lizard, and that was because the owner tipped the glass tank over when trying to clean it, and knocked it to the floor. The lizard was so dazed it just sat on the floor shaking its head until the owner picked it up again and put it in another tank.

So could these big lizards possibly have been escapees? I doubt it very much. There might have been one to start it all off, but the rest is probably some sort of wistful thinking or a kind of weak collective hysteria or some sort of need to be scared a little bit, like when people go on a roller-coaster or something similar. If you work yourself into a form of mild hysteria, I am quite certain even a belligerent common lizard could be reported as a ferocious monster ten times bigger than normal.

Ssssnnakessss!
And then of course we do have the escaped pythons and suchlike. Most of these cases are from bigger cities where various snaked go on a walkabout and turn up in the neighbour's kitchen or somewhere similar. But a few disappear into legends along the lines of alligators living in the sewers of New York. The Swedish equivalent is pythons living in the canals of Göteborg, the second largest city in Sweden. I have been told countless stories about these animals, but funnily enough, I have never been able to find a primary eyewitness, it is always something somebody has heard from somewhere else. More or less the same thing can be found in Malmö, another large Swedish city with its own system of canals. In this case it is supposed to be ferocious schools of piranhas!! Or maybe not...

If one ventures out in the wilds of Sweden, there are snakes out there as well, but they are in quite a different league. Some of them are probably not even snakes. There are quite a number of stories, especially if you go back about a century or so, about people meeting large, 2 or 3 metre long, black, and very slippery snakes, typically in fields near water. These animals have in some areas been considered quite dangerous. In a very famous Swedish novel "Bombi Bitt och jag" there is even a description of a fight between one of the main characters and one of these strange snakes. Actually I think these snakes are in fact big old eels that have been trying to get from one lake to another or perhaps to a stream of some description. This could possibly be of relevance in connection with the theory of sea-serpents being in fact big old sterile eels that haven't swam of for the Sargasso Sea, but have stayed in their pond or stream, and have just been growing, growing and growing – and eels can get quite old, so they will have plenty of time to get big. At the present date you rarely hear anything about big eels, but that is probably because of the sad fact, that the European eel is getting more and more rare, and nobody is exactly sure why.

The wheelies
If we put eels aside for the moment, there are still some stories about snakes that must be something else, if they are indeed real at all. And that is the wheelsnakes, very long animals, sometimes more than 5 metres, dark, and thick as a man's thigh. They

are or have been common in Småland, in south central Sweden, around ponds and lakes. They are very aggressive, and should be treated with the outmost respect, and then they of course have the rather strange ability – known from snakes on other parts of the world as well – of being capable of biting their own tails, thus forming a circle or a wheel, and rolling down hills at great speed when pursuing their prey or trying to get away from enemies.

Snakes with their own tails in their mouths, are known from large parts of the world in various forms. They are also known from quite a long way back in history. They are generally known as ouroboros, from an old greek word meaning "tail-devouring snake". The oldest forms of the ouroboros are from ancient Egypt, men the ancient greeks were fans of it as well, it is known form the European Middle Ages, it is used in various forms of yoga, in alchemy and in chemistry, so it is a fairly well-travelled and well-used snake.

In most ancient texts there are no indications of the ouroboros being an actual physical animal. It is meant as a symbol, meaning things like eternity, the life-circle, rejuvenation, eternal youth and various other things. They can also be a symbol of destruction – especially in Scandinavian folklore and mysteries, where the tail-biting snake sometimes have truly monstrous proportions. The biggest one of the lot is of course Jormungandr, the Midgaardworm, which is so big it encircles the entire Earth and bites its own tail, except at Ragnarok, the end of the world, when it lets go and starts wreaking havoc.

Slightly smaller are the special Scandinavian dragons, the lindworms, they are also snakelike, but quite a bit smaller than Jormungandr. Lindworms have a nasty tendency to make a nuisance of themselves around churches. They don't like them, and do their best to grown so large that they can completely encircle the church. Should they get big enough to be able to bite their own tail, they will most certainly do so, and crush the church like an empty paper bag. Lindworms by the way apparently used to be frightfully common in Sweden. As late as the end of the 19th century, there were plenty of sightings and worms making a nuisance of themselves all over the country. A true hotbed of lindworm activity was the area around Husaby, a town on the southeast coast of Vättern, the biggest lake in Sweden. This is perhaps not surprising, as lindworms are partly aquatic, as one can read about, should one wish, in "Weird Waters", my first CFZ-book.

At Husaby though – in 1869, 1878 and 1883 they were mainly causing trouble on land. People saw the giant tracks formed when the worms dragged themselves up from the lake to hunt for livestock and the occasional human. A number of people claimed to have seen the mighty beasts slithering across the landscape, and one unhappy farmer claimed to have seen one half of one of his cows protruding from the jaws of a lindworm as it turned and disappeared from his field. The rest of his cows by the way, were huddled in one corner of the field, as far away from the lindworm as they could possible get (and who can blame them). They were so stressed out by the experience that it took several days before they started giving milk again.

Actual dragons, you know the type with legs and wings are not especially common in Scandinavia, for some reason there are a lot more of them in the Baltic countries, but we will come to them in due course.

I offer no explanation for the existence of the various snakelike reptiles and their strange habit of biting their own tail. It is of course a biologically deeply unsound habit, although you do sometime see photos claiming to show exactly that, but none of these has as far as I know stood up to closer scrutiny.

A double-header
Sweden is also home to a rather strange lizardlike or perhaps newtlike animal, sometimes called the amphisbaena and sometimes various local Swedish names meaning things like "double-header", "two-header" or similar. Its most discerning

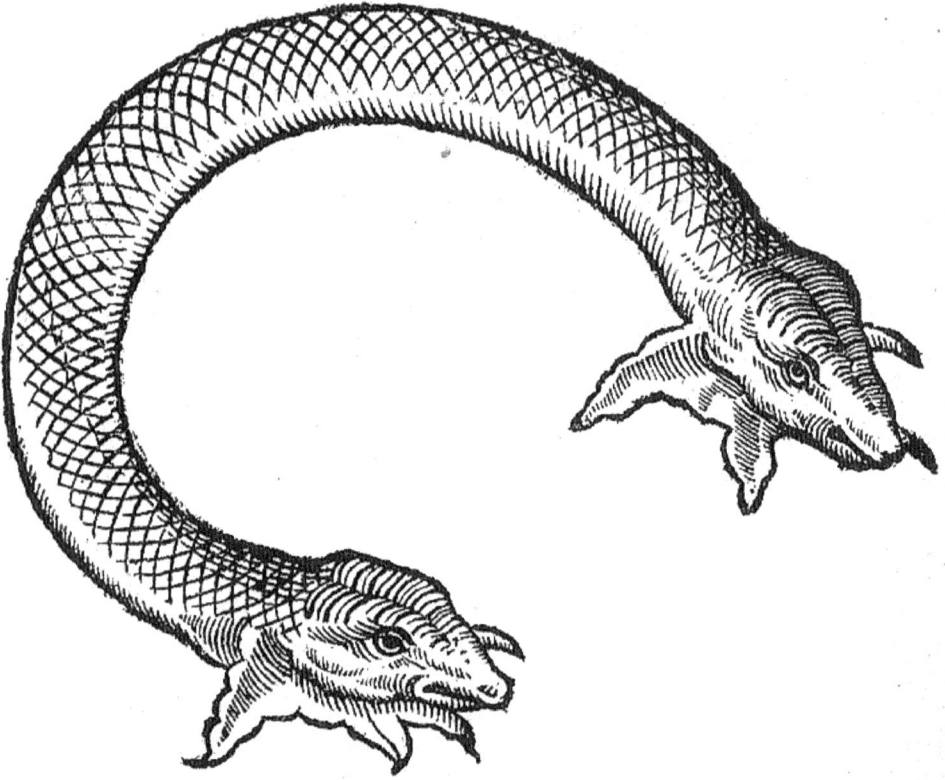

characteristic is that is has a head in both ends, or at least it looks like it has. This makes me think of the alpine tatzelwurm, a similar animal that allegedly lives in the alpine regions in Southern Europe (see also the chapter on Denmark). The tatzelwurm is a stubby animal with short legs , supposedly capable of making long jumps, with a wide and blunt head, and a tail looking somewhat similar. There is a rather famous photo of a tatzelwurm in existence (the only known one) – it was taken in 1932, but today most cryptozoologists agree that it is in fact a hoax. It is probably a picture of a wooden figure. You can actually still buy carved wooden effigies of tatzelwurms in some parts of Austria and Switzerland.

The Swedish version is not as well known, and it is in some ways quite different. It is not an animal of the high mountains, but instead to be found in lowland areas, in open forests or rocky fields. It doesn't jump, but because of its two heads, it can move backwards and forwards with equal ease. According to legend, it is also for some reason the first animal to wake from hibernations each year in the spring. It spends the winter in a hollow or a hole in the ground, and shows itself with the first warming rays of the spring sun. If you are a bit lucky, you can in fact see just the head poking out, while the animal is making sure there is no danger. In the summer you rarely see the animal as it is extremely shy and vary. It is neither dangerous nor aggressive, so in general humans and amphisbaena leave each other well alone. The stories are a bit vague as to what the animal actually look like. I have only been able to find one of two references to legs. In most cases the amphisbaena is described as a short, thick snake with a blunt head in both ends, usually dark brown or similar earthy tones.

I have never been able to find an eyewitness to an actual sighting of the amphisbaena, but I have met several people who seems to know a lot about the animals, which is a bit strange, but it is seems to have a permanent place in Swedish folklore as an animal everybody have heard off, but nobody have seen it.

What to make of this weird animal? It is in fact a bit difficult, not the least because there are a number of existing animals called amphisbaenids, looking exactly like the descriptions of the Swedish amphisbaena, i.e. a short, thick snake with a blunt head and similar tail. That in itself is probably a form of defence, as a predator would concentrate at least some of its attacks on the tail, which presumably can handle a beating better than the head. The major problem is that as far as everybody knows, amphisbaenians only live in South America, Subsaharan Africa and (one single species) around the Mediterranean, quite a fair bit from Sweden.

As I see it, there are two possible solutions to this little enigma. Seeing that the stories of the Swedish amphisbaena is concentrated in the south of the country, there is a faint possibility that Sweden was once or possibly still is, the home of a hitherto undescribed species of Amphisbaenids. The other possibility is, that stories about real life amphisbaenids – and some of these are quite old, and describes the animals as having two heads, somehow have made their way into Sweden, and have spread from there, people assuming that animals like that were also found in Sweden. And possibly the stories have been augmented by sightings of big old adders coming out of hibernations in spring. They are some of the first out of bed so to speak, they have rather big heads, and as they are capable of withdrawing into their holes quite easily and quickly, it would presumably be easy to conclude that this animal capable of moving back and forth with apparent ease, must have a head in either end, i.e. be the famous amphisbaena about whom one had heard many strange tales.

The other people
I know I have already treated trolls in the chapter on Norway, but Sweden is also home to manlike beings that are clearly not trolls, but something else, either very apelike or very humanlike, and in some cases also rather big. Here be giants so to speak... although when you go to the cathedral of Lund in southern Sweden, where

you can see the petrified remains of the giant Finn in the crypt below the cathedral, you are rather underwhelmed, as he is extremely un-giantlike. According to local legend, Finn was so annoyed when people started building the cathedral, and especially when they started ringing the bells, that he tried to pull down the entire structure by going down into the crypt and trying to destroy the columns carrying the main building of the church. Unfortunately for him, the instant he touched the first column, he was petrified and became a part of the actual structure he hated so much – sort of a fitting end I suppose.

Oh yes – there is also a story about why there are relatively few trolls in Sweden compared to Norway. That's simply because all the Swedish trolls upped sticks and moved to Norway when Christianity started to come to Scandinavia. Apparently the Norwegians took considerably longer to adapt the new belief than the Swedes, thus making living there more tolerable for the trolls.

The sexy skogsrå and others...
In the 1600's a number of rather weird court-cases took place in Sweden. Various

people, men as well as women were being accused of having sexual relationships with various forms of strange beings, mostly men with wild women or women with elves. One of them, Karin Svensdotter who was accused in 1656, declared that she had had regular sexual encounters with the king of the elves, and that she had given birth to seven of his children. The court decided that the Devil had confused her, and she was condemned to carry a silver cross at all times, which apparently put a stop to that affair. But in many cases, things did not end as well. In 1658 a young man who were wanted for a number of robberies, said that he had been living in the forests for a while with a skogsrå (a wild woman), and that she had been very jealous, but also very generous with her favours. That little story gave him a death sentence.

Whether these stories can actually be compared to stories of manlike beasts from others parts of the world is difficult to ascertain. Many of the culprits claimed that the various wild people were very humanlike, so they were probably nothing like Bigfoot or the Yeti, but they may have had some connection to the Asian almas and almastys, and especially the theory of them being surviving Neanderthals or some other early relation to modern humans. Or maybe they were just stories told to hide the fact that somebody was getting a little on the side.

Apart from these stories, Sweden can also boast a folkloric tradition telling about *vätterne*, an ancient tribe that had been living in Sweden a long time before modern man. They bred various animals, and preferred to keep to themselves. But as modern man moved in, they gradually disappeared, although some people claimed to see or meet them in increasingly more remote locations. It is easy to interpret these stories as a folkloric description of the meeting between Neanderthals and Homo sapiens. I wouldn't dream of suggesting that this is actually what it is, but it is worth thinking about.

The big smelly ape
Sightings along the lines of modern day Bigfoot sightings or some of the alleged sightings of big hairy apes from the U.K. in modern times are quite rare in Sweden. Swedish cryptozoologist Jan-Ove Sundberg have collected a number of sightings, but as I have already discussed in my book "Wild Waters", I am reluctant to use any of his material, as he has a tendency and a rather well earned reputation for being not as scientific as one could wish for. But I have spoken to a few eyewitnesses who claim to have seen something out of the ordinary that at one time frightened the life out of them.

The Bollnäs baboon... or whatever it was
In the summer of 1985 there were a number of sightings at Bollnäs about 300 km north of Stockholm, of a large, hairy, apelike being with a serious b.o. problem. The more sensationalistic part of the press claimed that the ape especially liked to watch young teenage-girls skinny-dipping in the local lakes, which probably sold more newspapers, but I haven't been able to find anything along the lines of searching for young teenage boys with a bit too much time on their hands.

What is considerably more interesting is the fact that several people in the area claimed to have seen this apelike thing late at night whilst driving through local forested areas, sometimes watching livestock from the edge of a field and suchlike activities. There was no actual attempt to search for the creature or indeed catch it, but whatever it was, it disappeared as quickly as it turned up. A real animal? A prankster? A spiritform – zooform phenomenon? Or a little bit of all of it. I don't know. I prefer solid physical evidence – just a tiny hair is all I ask.

The little people

Sweden is also home to a number of what one might call the little people – tiny humanlike or gnomelike or indeed sometime animallike creatures that lives under ground – but so is Finland, and again to avoid repeating myself to much, all of these will be treated in the following chapter, where I will take a closer look on Finland.

FINLAND - The land of the little people

According to Finland's neighbouring countries in Scandinavia it is a dull, dark and cold place, and the national pastime is suicide. The Finns are supposedly prone to drinking, knife fights and morose musings on just about all things in slow ponderous tones. And yet the country is also home to some of the world's most celebrated composers, architects and designers. The capital of Helsinki is a light and airy city, and in the middle of summer a place of outstanding beauty. A walk along the harbour front at the height of summer takes some beating. I grant you the winters are long, cold, dark and brutal, but on the other hand, if you can survive those, you emerge as tough, resourceful and a force to be reckoned with, as the Soviet Union discovered to their cost during the Second World War, when the tiny Finnish army gave the mighty Red Army a very hard time before they had to surrender. You can also see and hear a clear eastern influence in the country. It has after all, at various points in time been part of the Russian/Soviet Empire. That calls for some interesting mixes in beliefs, culture and folklore.

So all in all, Finland has a rather unfair reputation, although of course the Monty Python gang tried to remedy this deplorable situation with their "Finland, Finland, Finland" song. It is an exceedingly beautiful country with thousands upon thousands of lakes and vast forests, sadly also with a population of mosquitoes and suchlike stinging pests that should probably be counted in the high gazillions. But fear not – one can buy industrial strength repellent capable of stripping the paint of houses and clean chrome. Whether it actually works on the mosquitoes is a case for debate, but it smells and looks efficient. This rather large wilderness factor is to some extent also what makes it possible to support a large and varied fauna of strange creatures, the forests are deep and dark, most of the population is concentrated in the big cities along the south coast, and there are so many mosquitoes in the rest of the country, that not many people venture out into the wilderness. When they do, they see strange things – some scarier than others, but all rather weird. With so many lakes, there are of course plenty of aquatic creatures, but they have already been covered in my other book Weird Waters. Also the closeness to the vast Russian wilderness produces a large number of sightings of what must be wild animals like bears, wolves and so forth roaming across from Russia. Those I do not plan to cover here, as the explanation of those are fairly straight forward.

In keeping with the style in the rest of the book, I will also be covering a special subject in this chapter which could equally well be covered in all the others, but to avoid needless reputation, I will make a special study of the what one would probably call "The Little People" here, as the belief in those are pretty widespread in Finland. They can also be found in the various other countries, but the belief and interest in them is more located. In Denmark belief in the little people in modern times is almost exclusively found on the easternmost Island of Bornholm. In Sweden, the little people are more common in the south, whereas they are a bit rarer in general in Norway. There is some belief in similar creatures in the Baltic countries, but it is considerably more difficult to dig out information about them, as there is a considerable language barrier involved here, but more on this later on. To some extent there is also a language barrier in Finland, although not as pronounced. The Finnish language is extremely complicated and difficult to learn, but luckily a large proportion of Finns speak Swedish and/or excellent English, so one can usually get by, but in more rural areas, especially up north, they either can't, or won't speak anything but Finnish.

To some extent it is a matter of national pride, as one can find in so many other countries – people of Catalonia who refuses to speak anything but Catalans, some areas of Wales where people refuse to speak anything but Welsh, and so on and so forth. And of course there are people who couldn't care less whether tourist, outsiders, foreigners or whatever you choose to call them, can understand them or not. One rather arrogant young Finnish woman once told me, that the only reason she spoke English was because it was part of the curriculum at her school, otherwise she saw absolutely no reason to speak anything but Finnish or be interested in anything that happened outside of Finland, except when it had something to do with other Finns. I should hasten to add, that she was a rather extreme exception. All other Finns I have talked to have been nice and friendly, but perhaps a little cautious.

For some reason which I haven't quite been able to fathom Finland is also home to considerably more alien big cats than the rest of Scandinavia. This might be because the country has a bigger population of lynxes than any of the others, but it might also be because more people have big cats as pets – what else is there to do during the winter? Or it might for some other exotic reason that escapes me, I don't quite know.

But big cats are being seen on a fairly regular basis, although how they survive the Finnish winter, unless of course they are all Siberian tigers, I have no idea.

Besides that, one can meet a sprinkling of various others weird beasties, some of which I will take a closer look at. There are also a few very, very strange ones, things that have only been seen once or maybe twice, and that falls far outside the realm of any of the typical groupings of weird creatures. I will present a selection of these, but I will not try to analyse them in any special way, as the number of sightings is frustratingly small and also because they might be nothing more than urban legends in the making. We shall see.

The little and the slightly larger people

For some strange reason I have never quite been able to figure out, Finland is home to a bewildering variety of little people, running around on, and indeed below ground as it were. Some of them are even capable of flight, as they sport a rather fetching set of wings. I have no idea why the belief in the little people is still so rife in Finland. Only Icelanders are bigger believers as I have demonstrated in the chapter on Iceland. In the rest of the area there are small pockets where the little people are still very much a part of people's everyday life and beliefs, such as the island of Bornholm in Denmark, where a Danish researcher in 2014 actually got a 3-year grant to study the beliefs in the little people and how it had affected modern day society.

The belief in the little people is quite straightforward and not, shall we say, infused with anything spiritual like so many new age religions have tried to install into the little people. In some areas the beliefs have been heavily commercialized, such as the theme/ amusements park Brændesgårdshaven on Bornholm where everything is centred around the little people or jobos as they are called here. But by and large the little people are just something that are there, like foxes, badgers and other things you may encounter on a walk in the local woods.

The Christmas people

One group of the little people are very much in the forefront of just about everybody's lives in the Scandiniavian countries around Christmas time. Those are the "nisser" as they are called in Denmark and Norway, or "tomter" in Sweden. They are called tonttu in Finland. Their origin lies way back in history, when they started out as a kind of house-gods or house-protectors, but in the late 1800's they suddenly through the medium of various stories and novels became irreversibly linked with Christmas by being promoted to help Santa Claus with everything from making Christmas presents to well-behaved kids, to mucking out the stables of the reindeer. The rest of the year they presumably go back to their normal life in houses, forests and mountains.

Oh yes, a little aside here. The actual location of the house/workshop/castle of Santa Claus have been the cause of an almost perennial diplomatic crisis in Scandinavia. According to the Danes, Santa Claus has his home in Greenland, but according to the Finns, he lives in Finland, so around Christmas time, relations between the Danes and the Finns usually get a little strained.

"Nisser" look nothing like the elves that traditionally help Santa Claus in places like the UK and USA. They are much closer related to the Irish leprechauns. They look like tiny humans, but dress in grey or brown, it's only around Christmas they put on what must basically be called a work uniform, i.e. red cap and/or red clothes.

If we go a couple of centuries back in time, sightings of "nisser" were fairly common all over Scandinavia, but today they are apparently extremely rare. Danish author Herman Stilling (1925-1996) claimed to have seen a "nisse" sitting on a fencepost, but that it disappeared as soon as it realized he was looking at it. Another sighting – this time from Sweden was reported by a girl in 1978 who claimed to have seen a "nisse" stealing a bun from her family's picnic table outside of Helsingborg during the summer holidays.

Strange happenings at Frostens Å (The Frost Creek) and a few other places

Frostens Å can be found on the island of Funen in Central Denmark, and for some reason it is a place where children have a lot of weird sightings, but adults never see anything – they are probably too busy thinking of mortgages and similar boring stuff. The stories can be traced back several hundred years, so the details are scant, but usually they tell about children who either alone or when in the company of perhaps

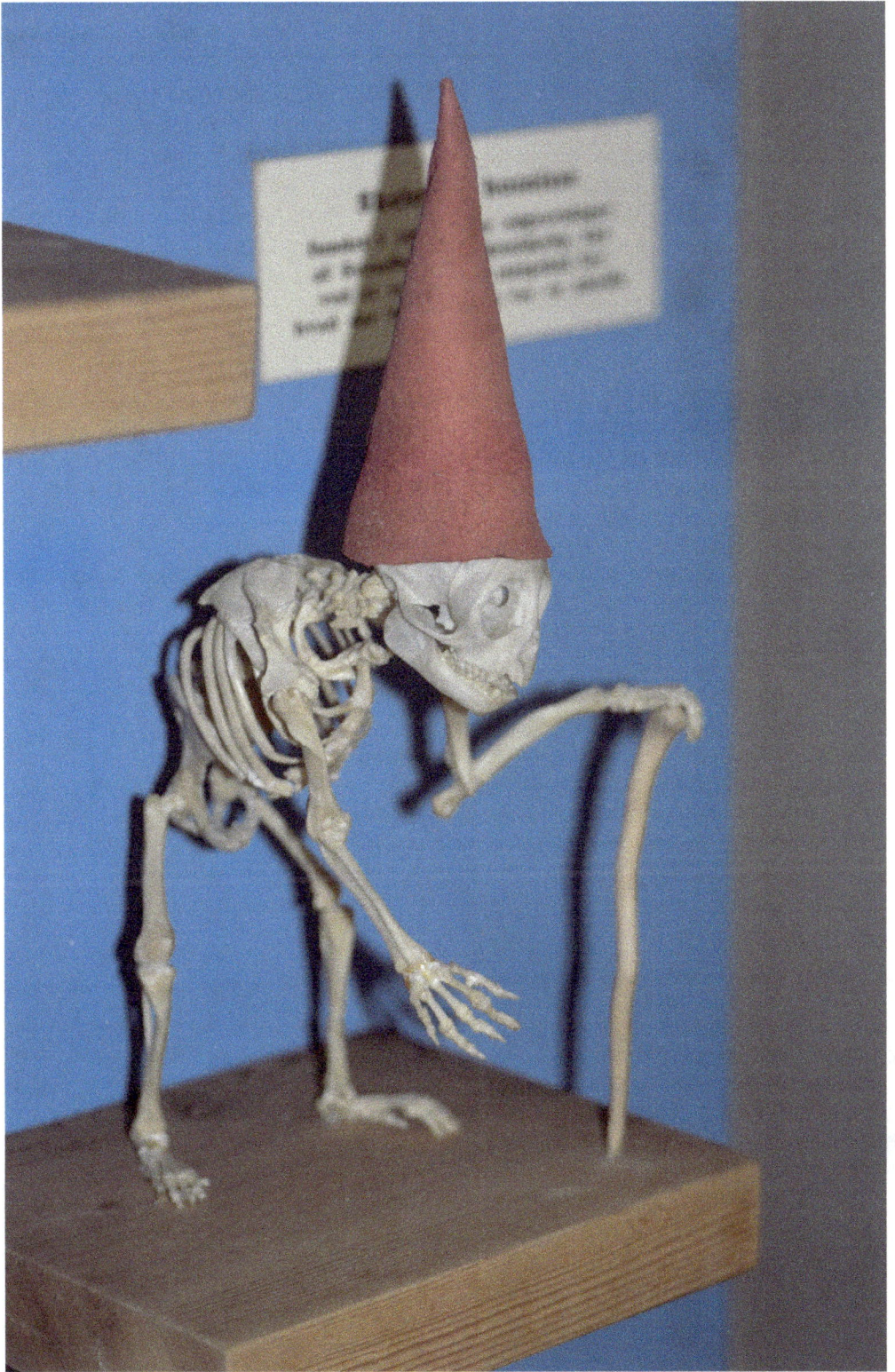

their parent, see tiny men dressed in grey. They are usually alone, and can be standing at the banks of the creek looking on the people walking by, or perhaps they are sitting on a tree-stump, a fencepost or on a big rock, keeping an eye on things.

Today the sightings are few and far between, but I have managed to talk to a couple of people who have experienced strange things. Nisser are known to be basically very peaceful, but also eager practical jokers. One man who has been at the receiving end of a nissejoke, was out mushrooming in Langesø Forest close to Frostens Å in the autumn of 1988. After a couple of hours he decided to take a break on a fallen treetrunk near the creek. He sat for 10-15 minutes smoking his pipe, and then started on his way back to his car. He had gone about 200 metres when he realized he had left his pipe on the tree-trunk and hurried back to find it. He found it quite easily – only thing was, it was now lying on a tree-stump on the other side of the creek. And, as he pointed out to me rather forcefully, when he told the story, there was at least a kilometre to the nearest crossing point.

Another sighting comes from a woman who in 1981 was a 6-year old girl on a picnic with her parents. Things started to get weird, when a thermos filled with hot coffee suddenly disappeared.

> *"I remember quite clearly how irritated my father was, when I kept insisting, that the thermos could be found in a tree about 30 metres from our picnic, and that I had in fact seen a small man dressed in grey put it up there. In the end they were so desperate that my father did in fact searched the tree and found the thermos. My parents plain and simply refused to believe my story and said I had hid it. But I didn't. I was afraid of heights – I still am. It was a small man. He was about my size, and he had a beard. And I am certain he actually winked at me, when he took the thermos."*

The tonttu

The Finnish versions of nisser, the tonttu, are very similar to their relatives in the other Scandinavian countries, but they do tend to exude a slight sense of menace. Not that the "nisser" couldn't be problematic, far from it. In the olden days they had a rather nasty habit of setting fire to whatever farm they were living on, if the humans did not treat them with suitable respect – an act rather euphemistically called "Making the red cock crow"!

The tonttu are small, but much more childlike than the "nisser", and they usually also have pointy ears. Just as their colleagues further west, they are mainly active in the world of humans around Christmas-time, when they work for Santa Claus. Not in a

practical sense mind you, but by evaluating the behaviour of children. They will report any transgression of good behaviour to Santa himself, and that can have serious repercussions for naughty children, some of which have been known to disappear without a trace. Apart from that, the tonttu behave more or less like their relatives and are usually closely associated with human habitations, not just farmhouses and stables, but also – we are in Finland after all – with saunas. In fact there is a special breed of tonttu, the sauna tonttu, which must always be treated respectfully. In fact the tonttu actually demands a good deal of respect – and a permanent supply of food, preferably porridge or soup. But if you meet their demands, they are friendly and good-natured.

Sightings of the tonttu are few and far between. Like so many of the little people, they are usually only seen by children, and even then only glimpsed out of the corner of an eye. There are almost no details in the sightings I have heard of, but just to give you an idea of a typical sighting:

In 1963, three brothers aged 8, 10 and 13, was playing in their room in their family's house at the southern coast of Finland. It was late December, but as it was midday, so there was a fair amount of sunshine. They boys were very noisy and making a mess of themselves and their rooms. They had been playing more of less randomly for more than an hour, when they accidently all three happened to end up in the middle of the room with their backs to the window. And then, according to Mika, the oldest brother, things took a turn for the strange.

> *"We were just standing their laughing, when all of us suddenly remembered a western movie we had seen a few days earlier. So we all looked at each other, laughed, turned towards the window and yelled: Hand's up stranger!! Imagine our surprise, when we saw, although only for a fraction of a second, what looked like two small boys with red caps, who had been sitting on the fence looking at us. They disappeared so quickly, like somebody turning off a lamp that I thought I had imagined them, but it turned out my brothers had seen them too. We never told our parents, we were certain they wouldn't believe us."*

What did they see? A couple of squirrels perhaps? I don't know, but when I talked to Mika 40 years later, now a very serious schoolteacher working in Sweden, he was adamant, that he and his brothers had seen two tonttu.

The menninkäinen
The menninkäinen are closely related to the tonttu, but there are some differences. They look roughly similar. They are small and basically human like, but have fairly large heads with pointy chins, pointy noses and small beady eyes. They have been seen to wear pointy hats as well, but apart from that, not much is known about their clothing and general habitats. This is probably partly because they are nocturnal,

rather timid and live under the ground in remote forest areas, although the more curious among them sometimes approaches human settlements. They are usually responsible for items going missing in rather unexplainable ways or suddenly displacing themselves, but in general, if you want to see the menninkäinen, you have to meet them in their natural habitat – the forest.

At nigh the menninkäinen er almost impossible to see or find, unless you happen to stumble into one of their parties. They are rather found of music, dancing and merriment, and in those cases have a tendency to forget their shyness and let their hair down as it were. One does have to be a bit careful around parties when they are being held by the little people. Times has a different meaning to them – a little people minut can easily turn out to be 10 years in human terms, so you might find things have changed a bit when you return home from one of those parties. Occasionally you do sometimes meet menninkäinen in broad daylight, and if you approach them carefully, they are usually quite friendly, although you have to be careful if you f. instance are wearing any kind of jewellery. They are very partial to shiny objects, especially rings. (Traces of Gollum in here somewhere?)

You do have to be careful though around them though – because if you are misbehaving, disrespectful to the forest or just being loud and obnoxious – especially if you are a child, you might find yourself being led seriously astray, and then abandoned somewhere deep in the forest, left to your own devices to find your way home.

I have never been able to find anybody, Finn or otherwise, who have claimed to have met one of the menninkâinen, so I have no theories to offer as to their possible existence or whatever creature that have inspired to these stories, but since they can lead you astray, a potentially lethal situation, I am inclined to believe, that they are in fact small and relatively benign versions of the bogeyman – designed to keep people from getting lost in the wilderness, or at least not put themselves in situations where they could get lost. If you are keeping an eye out for menninkäinen, chances are that you are also taking notice of where you are, which way you are going, and hence will not get lost. But then again, what do I know – I have a terrible sense of direction, and is quite incapable of finding my way without a map, so I would be utterly doomed, should I chance to make the menninkâinen angry.

The keiju

With the keiju we come to the very smallest of the little people. The descriptions of these vary quite a lot. They vary in size from a 10 year old child all the way down to something only a couple of centimetres in height. But a common theme for all of them is the fact that they look like very beautiful, slim and elegant humans sporting a set of rather elegant wings, either in the shape of a set of fairly colourful butterfly wings, or a set of dragonfly-like wings looking very much like stained glass windows. Should one be of a zoological mind, one could conclude that we must be dealing with a least two different species on account of their dissimilar wings.

The keiju prefer to stay well clear of humans, but they are quite friendly should you chance upon them. They will smile politely, but the chances of them inviting you home is remote. By and large they consider humans to be gross and coarse creatures with as much style and poetical sense as a flock of mud-covered water buffaloes. They much prefer to live in peace and harmony with all the things around them. Slightly hippiesque, but not a bad lifestyle all things considered.

You won't find many people who claim to have seen keiju at their maximum size, or largest manifestations if you like. Apparently they can also change size, and the maximum setting is only for very special occasions. But there are quite a few people who think they have seen them at their most minute – indeed I have even been taken out to watch some of their gatherings, and have seen quite a number of them – or at least that's what my guides have told me. It is not as strange as it may sound, especially if you consider the preferred habitat of the keiju – moist meadows and areas around ponds and slow moving streams. You can probably do the same almost anywhere if you visit similar areas in late spring or early summer.

Sometimes there is just one or two, and sometimes you can see them by the hundreds, tiny humanlike figures with long slender legs and arms and gossamer thin wings, all fluttering over reed beds and shallow areas. They look quite fantastic against a low sun that turns them golden or silvery. They are of course not, I hasten to add, humanlike in any way. They are most likely – or at least they were in all the cases I have seen them - swarming mayflies. Beautiful and rather weird looking insects, but neither magic nor supernatural, although some of the people I have presented this too, have been considerably less than willing to believe me, because they do look like extremely slim and tiny humans. One even suggested I was making a fool out of them. I had to catch a couple to convince them, but that only made them certain, that the real keiju would hide themselves among the insects…

Fauna ignorata

In his book *"Fauna Ignorata"*, Danish author Michael Lycke have suggested another and totally different explanation for the various forms of little people that we find in Scandinavia, and especially in Finland. I am not quite certain what to make of his book, as it is written I believe, with tongue very firmly in cheek, but it is a good read, and very much worth a closer look – and why not?

Lyckes idea is that the various stories of the little people, nisser, leprechauns or whatever you prefer to call them, are not products of people's imagination mixed with folkloristic traditions, but actual sightings of rare, but existing animals – not in any way related to humans, although they may look like it. According to him the Finnish tonttu for example is in actual fact a small but formidable predator capable of killing and eating a wolverine if need be, and all the stories about friendly little humanoids have been made up following a sort of unwritten consensus to hide the existence of these rare but extremely dangerous animals. Lycke also suggests that things like fairies are in fact small vinged reptiles, and not small winged humans with a pearly laughter and a penchant for dancing and partying. He describes no less than 9 different species.

The book is of course to some extent a joke, but perhaps not all the way through. It is written with dedication and attention to detail, and it gives if nothing else, a good idea of how animals can pass from reality to legend and back again.

The Natural History Museum in Copenhagen does in fact take this seriously (sort of). For many years they used to have a special exhibition around Christmas time, concerning the biology of nisser and the research into their biology. They even used to exhibit a nisse-skeleton, allegedly found in a posthole during an archaeological excavation in the town of Svendborg in central Denmark in the 1970's. Or maybe it is nothing more than a bit of Christmassy fun, and the skeleton is made from various bits and pieces and put together by one of the museum's talented taxidermists.

Beware of the peikko and the hiisi

From the small, usually friendly and nice, we move on to the slightly bigger and a lot less nice. There are plenty of these as well, so I won't go into too many details regarding the peikko, as they are usually known. Some of them are big enough to turn into landscape features when they fall asleep – something they occasionally do for centuries, but there are also smaller examples, and some are even invisible, such as the hammaspeikko who makes holes in children's teeth, and there is of course also the känkkärankkåa, that gets children to misbehave.

From a cryptozoological point of view, by far the most interesting of the lot are the

hiisi. They are very far from benign. They are humanlike, but very wild and primitive – considerably more hairy than normal humans, but not completely ape-like, and they have a disconcerting tendency to attack unwary travellers in remote areas. To some extent they have troll-like characteristics, but are also reminiscent of very primitive humanlike beings. For some reason they are especially common, or less rare, along the Russian border. As a general rule, the Scandinavians do not have much in the way of sightings of things like Bigfoot or other snowman-like creatures, but around here they do.

The number of sightings I have been able to unearth is probably only a fraction of the actual number. Most of the sightings are more than 50 years old, with a peak around World War II, but that means they can only be found in Finnish literature or newspapers – and that effectively places it out of my reach. Finnish is not an easy language to learn if you are not Finnish yourself. But fear not, there are newer sightings as well, and sometimes you are lucky enough to talk to people who can remember stories their ancestors have told them. I have been able to gather around 30 sightings stretching over roughly 150 years, but I will only give some examples here.

The creatures people see around here are not as animallike as some of the other man-beasts of the world. They are more humanlike, and seem to be closely related to the almas of the Kaukasus area and the almasty from Siberia. They seem to have some form of primitive society and/or language, and more than one witness have described them as primitive humans, "like something from King Kong or King Solomon's Mines". The most common name for these creatures I have heard is the hiisi, but a couple of my witnesses have called them almasy and almastys as well, so they might in fact be the same types of creatures.

War stories
One of my late father's colleagues (he was a violinist, playing in the symphony

orchestra in Malmö in Sweden) was a Finnish flute-player, who in his youth has fought the Red Army in the snow-covered forests of Eastern Finland. According to him it was common knowledge among the soldiers that you not only ran the risk of having a run in with Soviet soldiers, when you were out and about, but also with hiisi. The fins would send out small patrols on ski to keep a lookout for the Red Army, and although the soviets rarely saw them, the hiisi did. They were under constant supervision, and only if it came to actual battles or shootings, the hiisi would disappear and melt into the trees again. Some of his friends in the army claimed that at least one of the hiisi had been found wounded after a battle, but as they soldiers had more than enough to do with taking care of their own dead and wounded, they had left it where they had found it. It was the size of a powerfully built man, but extremely hairy. Some versions of the story claims the creature was naked, whereas others state that it was dressed in what looked like primitive clothes made of animal skins. He had no idea whether these stories were true, or whether the more seasoned veterans were just trying to wind him up, but although he never said anything about actually having seen any of the hiisi, he did say, that he always had the distinct impression of never being alone anywhere in the forests.

My grandfather also had a friend who had fought in Finland against the Red Army. He had volunteered to join the Finns, and although he didn't like to talk about the war – he had seen far too many atrocities, and I was only 14 when he died - but he did tell me, that he had not only heard similar stories about hiisi, or wild men of the forest, when he was in Finland. But he had actually seen one of them. Late one night, when he was on guard duty, he suddenly realized he wasn't alone anymore. A figure had suddenly materialized next to a big tree only about 20 metres away, and was just standing there looking at him. He felt no threat of any kind, just a wave of curiosity. It was also very clear, that it was not a Soviet soldier, or even a local. It was a very big hairy man, with broad powerful shoulders. The most interesting thing was that he or it was carrying a long stick or possibly a spear.

Later stories
As soon as things started to calm down after the war, the number of sightings started to dwindle, and today sightings are considerably rarer. With the forests more tranquil, the hiisi probably withdrew to the remotest areas they could find, and now they are only seen occasionally, either because a rambler blunders out their way, or they see something that arouse their curiosity.

In 1997 a group of friends who were out on a moose hunt reported afterwards that they had all had sightings of what was variously described as a gorilla, a chimpanzee or a hairy human, watching them from a distance while they were all standing at their various posts waiting for a moose to show up. The man/animal or whatever it was, just seemed to be studying them out of curiosity, but did not attempt to get closer or to contact them in anyway, and he/it did not seem to be hostile. One of the hunters, a lawyer from Helsinki, the one who thought the creature looked like a hairy human, also said, that the creature seemed to be leaning on a long stick of some kind, perhaps a spear, but that he couldn't be certain, as the thing was standing among the trees on the other side of a clearing in the forest, at least 150 metres away.

In 2004, three tourists, two Danes and a Swede who had been birdwatching, walking and fishing, had found a great grey owl in a tree and had stopped to watch and photograph it. While they were there, a strange hooting sound, not coming from the owl, and not sounding like any owl species they knew, flushed the bird and made it fly away, and a few seconds later two figures came running through the forest with great speed.

> *"At first I thought it was a deer of some kind. It was very fast, bounding through the trees like they weren't there, but when it got a little closer it separated, and I realized it was two figures running very close together. It was difficult to see any details among the trees, although they were only about 50 metres away when they were closest, but it looked like a couple of humans of average size, apart from the fact that they looked very dark and hairy, as if they were both wearing heavy fur coats. I tried to take a picture of them, but there was nothing but trees on the picture when I studied it afterwards. We might have been hallucinating, but I don't think so. We all saw it, and we all thought it looked like running humans."*

Explanation, please!

There is not much to go on, when trying to identify who or what the hiisi are. There is of course a fair number of sightings, but virtually all of them, or at least the ones that I know of, are tantalizingly short, often only lasting a few seconds. In most cases people have also had quite a lot of difficulties getting a clear view of the creatures. Often the hiisi are only glimpsed out of the corner of one eye.

And when it comes to physical evidence, scarce doesn't even begun to describe it. I have heard a couple of stories about the finding of prints of naked feet, but have never been able to trace photographs, drawings or casts of them, and since they are described as just like human feet, they may be nothing more than the footprints of modern Finns out on a bare-footed walkabout. I have also talked to one man who claimed he found a hiisi spear in the forest near Turku when he was a boy and brought it home. Unfortunately his father didn't believe his story, got angry, broke the spear (or ruddy stick as he called it), and threw it in the fire. My witness nevertheless insists that it was a primitive fire-hardened spear, and that I smelled distinctly of sweat, when he found it.

I have never heard of anybody finding hairsamples or anything like that – and as the hiisi are described as being fairly hairy, it wouldn't be too farfetched, but nothing at all. It is a quite a shame, as it is perfectly possible to identify an animal from just a small hair-sample. You can't of course identify an unknown animal, since you have to compare the hairs to samples of a known origin, but you would be able to say, that a new species was out there, waiting for a more concentrated effort to find it. You could of course also do a DNA-analysis, but the same thing applies. You can only identify a species if you have something to compare it with. If you can't find a match, chances are you are dealing with a new species.

There is also a story of one of them being shot, but as the detail of the story varies quite a lot, depending on who's telling it, and the location varies wildly as well, it bears all the hallmarks of a modern day urban legend. Similar stories have been told from Russia, China, North America and several other places – and quite a lot of them have been shown to be hoaxes, especially stories about dead Bigfoots in North America.

So is it in fact possible to conclude anything at all? Not much to be honest, but here goes...

There are in fact quite a lot of theories, but none of them seems to stand out in any way. There have of course been sightings of similar beings from other parts of the world. They also vary quite a lot, from distinctly apelike, such as the Indonesian orang-pendek, to the equally distinctly humanlike such as the almas or almastys of Asia. The hiisi is obviously most similar to the latter, but that's about all we can say for certain.

One could argue that there are two different possibilities. One is, that they hiisi have some form of physical reality. The other one is that they don't!

There is no indication of the hiisi being apelike in any way. They look very much like primitive, almost prehistoric humans with the sticks or spears they carry around, and indeed their primitive clothing. But if so, what kind of primitive humans could they possibly be. Every now and then your hear stories of a tribe being found that never had any contact with the modern world before that, but that is usually somewhere deep in the Amazonain rainforest or in the mountains of Papua New Guinea. Surely not in Finland? Some have theorized that Asia is home to small populations of Neanderthal-like humans, and it is conceivable, that some of these, if they actually exist, might have spilled over into Finland. Others have suggested, that the hiisi are in fact primitive sami-like people or the ancestors of the modern sami that somehow have managed to live on, hidden from the modern world in wild and remote areas.

There is absolutely no physical evidence of the existence of the hiisi, whatever they might be, although some people claim that strangely shaped rockformations that can be seen in some areas, are in fact the work of the hiisi – a theory not, I might add, supported by modern science. So in fact, there is no way to decide. Neither of the theories is especially likely to be true.

On the other hand, the hiisi might well be some form of figments of people's imagination, or indeed images that occasionally slip through cracks in the space-time continuum, so people are actually seeing things that happened thousands of years ago. They are probably not frauds or anything like that, although some of the sightings can be misidentification of animals – there are for instance bears in Finland. Or they might be some form of distant memory, trigged by fleeting glimpses and movements in the forest, of a time where tribesmen were in fact walking these woods, and where knowledge of the could be vital for your survival. After all, the hiisi are supposed to be dangerous, and should only be approached with extreme cautions. They might be very interested in having you for dinner!

Here kitty…

Another strange sighting to end this part of the Finland, and one which give me an excellent lead-in to a another section of the book, comes from Kuhmo, about 500 km northeast of Helsinki, where in 2001, a man driving north along highway 75 towards Kuhmo, saw what he described as "a hairy marathonrunner". The rather strange expression comes from the fact that he saw a man-like figure running along the road, with the long easy stride of a long distance runner. He only saw him from the back, and not only did the runner seem to be naked, but he was also quite hairy and powerfully built. The sighting only lasted a couple of seconds, as the runner turned and disappeared into the woods as the car got closer, but the really weird thing was that alongside him, also running quickly along, was a big greyish brown cat!.

Cats of all sizes and shapes

For a country whose official list of cat-species is as short as they come – that is to say one – the European lynx, Finland has produced an awful lot of sightings of big cats. I have no idea why there are so many – in some years the country have even rivalled the UK for the sheer number of sightings – perhaps there is nothing to do during the long dark Finnish winters than breed big cats and let them run free, although I doubt it. But the fact is that cats of all sizes and shapes are galliwanting all over the Finnish landscape. According to some of the stories I have been told, they can even be seen in the centre of large cities brazenly walking down the middle of large roads, albeit in the middle of the night, although I have one sighting on file of an extremely large brown cat walking along the harbour front in Helsinki in broad daylight. This particular sighting is probably not worth much, as the witness claimed there was lot of other people about, but that none of them paid the cat the slightest bit of notice, and that he seemed to be the only one capable of seeing the animal. Now, if you'll pardon my expletives – what the xxxxxx is going on?

If you sift through the various sightings, it is possible to sort them into several very distinct groups, and I shall try to describe them and analyse them as such. The number of sightings is actually so large they could easily warrant a book of their own, but I think that should be a job for a Finnish author, so I will not try to analyze many single sightings, but just give what I consider good examples of the various groups.

Lynxes at large

The first group consist of sightings of medium sized to large cats with very long legs and no tail. They are usually described as being grey of greyish brown, sometimes

with weak spots and usually also as having very large feet. In some cases people describe large tufts of hair on the tips of the ears, but by no means in every one. The animals are extremely shy and nervous, and are usually seen far from any large city or town.

I have no doubts that these are sightings of Finlands only proper cat – the lynx. The sightings are all well within the parameters of how this animals looks, although a few of the stories contains some rather extraordinary elements, such as the sighting made by a young woman on a road north of Turko in 2001. She claimed the long-legged grey and tailless cat she saw run across the road was the size of a Sibirian tiger. That rather hefty piece of size exaggeration must surely have been caused by sheer fear. She did after all tell me she was so shocked she almost ran her car off the road.

A couple of other sightings are a bit more difficult to explain. In both cases the general look of the animals leave no doubt that they were in fact lynxes, but one animals seen in 1998 very close to the Swedish border near Tornia, was described as having a bright rusty-red coat, whereas the other animal seen just west of Rovaniemi in 1988 had a distinct bluish hue to its coat.

Now cats do sometimes turn up with very unusual colours. British cryptozoologist Karl Shuker have described several cases in his various books and articles about mysteries cats, and I have described a case of a green housecat in the chapter on Denmark in this book, so I am assuming, that these two aberrant lynxes are simply extreme colour variations.

Moggies and suchlike
The next group of cat sightings are fairly straightforward as well. It consists of sightings of small to medium-sized cats with an extremely great variation in patterns, thickness of coat and agility – or lack of same. There is as such no doubt in my mind that all of these are sightings of domestic cats running loose, even though a couple of them describe animals of a rather impressive size. I have been told stories about cats as big as Alsatians, and although breeds like the Maine Coon or the Norwegian Forest Cat can grow to rather hefty sizes, part of it must be caused by people being scared or startled and misjudging the size of the animals. The animals have been seen all over the country, from the middle of large cities to wild areas far from any kind of habitation. Especially the last fact has driven many people to state that it definitely wasn't domestic cats, because "you wouldn't find them out there". This is of course wrong – as any environmentalist will tell you. Cats are very adaptable creatures, capable of living in almost any environment and wreaking absolute havoc among native animals if they decide to settle in an area where they are not a natural part of the fauna, or if they escape from captivity. But for some strange reason, people seem genuinely surprised when they see cats under these circumstances.

Domestic cats have always been steeped in legend and superstition, and to a certain extent they still are. Many a time when I have tried to explain to people, that what they have seen have been an ordinary moggie, they may reluctantly concede to the cat id, but will inevitably tell me something along the lines of – "Yes, but this was no ordinary cat, because..." (insert any scary story here). This I think is something one has to be very much aware of when interwieving people about animals like this. Local beliefs and superstitions can very much colour people's perception of animals, According to some of the Finnish eye-witnesses I have talked to, domestic cats are f. inst. capable of killing fully grown reindeer by going straight for the jugular, and they can of course be seen as omens of just about anything bad and evil you care to suggest. For good measure I could throw in a couple of the stories I have heard, where people claim to know of domestic cats trying to kill babies or drag them out of their cribs – but I won't.

"Ordinary" big cats

This group of sightings describe what I would call "ordinary" big cats, that is big black, brown or spotted cats – typically leopard or small lion-sized. Sightings like these are well-known from many parts of the world, in Finland they are especially common in the southern parts of the country, but in actual fact the animals have been seen in almost every part except the very far north.

Just a few examples to give a general idea of the extent of the phenomenon:

Imatra in southeastern Finland, only about 10 km from the Russian border: Several sightings of a large golden-brown or brownish cat. The animal was fairly shy, but looked powerful and muscular, with short and rather stubby legs, a small head and a very long tail. In general fairly silent, apart from uttering a muted growl every now and then. At the same time as the sightings, several people heard a strange yowling sound in the same area, but the big cat was never seen uttering this sound. A puma perhaps? Sure sounds like it.

Kristinestad in western Finland at the coast of the Gulf of Bothnia: A very large golden-yellow cat with black spots was seen on several occasions. It was very muscular, with a big head, and a distinct affinity with water, as it was seen swimming on several occasions. One cannot help but wonder if this was in fact a jaguar, as they are fairly at home in water in the natural habitat in Central and South America – almost as much as the tiger.

Salo in Southern Finland: On a golf course a couple of kilometres outside of Salo, a small group of people saw a medium-sized cat with a very distinct brownish-golden marbled fur hanging upside down from a branch in a big tree, apparently trying to sneak closer to a small group of pigeons at the end of some branches. When the animal saw the people, it simply let go of the branch, fell to the ground and ran away. The rather special way of moving about in the tree makes me think of a clouded leopard, but again not the slightest evidence for its actual existence.

Helsinki, the capital of Finland: Various people living in one of the northern outskirts of the city claimed on several occasions to have seen a cat that looked exactly like a miniature version of a lion. It wasn't a cub, but clearly an adult animal, but only about half the size of a lion, or perhaps a little smaller, and considerably slimmer. The only cats I can think of that looks like miniature lions are the two species of golden cat. They can occasionally be found in captivity, but in general they are quite rare.

Vaasa in western Finland, close to the Gulf of Bothnia: Various sightings of a large black cat in the area. According to some of the descriptions I have seen, it could easily have been a panther or a black leopard if you like, but the animal was fairly small and long-legged, although considerably bigger than normal house-cats. We could of course be talking about a real leopard here – sizewise they do vary a lot in the wild depending on where and how they live, but in other cases housecats have been shown to grow excessively large in some places in the world, when they have been living in the wild for some time.

In the Vaasa case they story may be more complicated than you would think at first sight. According to some of the residents in the area, there are also stories about a ghostly black dog in the general area of the city. This dog – much along the same lines of other black dogs in other parts of the world – patrols certain areas, shows itself to people every now and then, although sometimes it is only seen rather than heard, and seems to be very much linked to the same specific areas. It is not inconceivable that the dog has morphed into a cat somewhere along the line, and as such probably have no physical reality, but belongs in the realm of ghosts and zooform phenomena.

The Elvi-case
The most famous alien big cat case in Finland is probably the Elvi-hunt in the summer of 1992. This all started on the 25th of June near Imatra, which is located in eastern Finland, quite close to the border with Russia. A forestry worker named Matti Arvinen claimed he suddenly came face to face with nothing less than a female lion close to a place where a half-eaten moose had been found a day or two before. Arvinen was, as one would imagine rather chocked, and a bit uncertain as to what he should do, but he reported the sighting to the authorities, and that would probably had been that, but two days later, the lion was seen again, this time in Kekälemäki about 30 km northwest of the original sighting, and on the 29th two girls swimming in a lake near Immola saw the animal as well.

And then all hell, or rather panic, broke loose.

The authorities, local as well as national started discussing the case, the media had a field day, and people in the area started to get a little vary of going out alone, especially after dark. And suddenly, as is often the case, people started seeing lions all over the place, some sightings even as far away as the extreme western coast of Finland, several hundred kilometres from the original sightings. It was round about this time, the lion was named Elvi, people started organizing hunting parties, and

various classical explanations was put forward, including "it must have come from a circus" and "it escaped from an overturned truck/traincar, when a circus had an accident on their way from one place of performance to another". Needless to say nobody bothered to check whether any accidents had been reported – but then again someone claimed it had all happened on the Russian side of the border. As far as I know nobody tried to contact the Russian authorities either. These kind of stories – upturned circustruck/train is known from all over the world, and has been cited as the cause of countless strange animals turning up in all kinds of places. I have actually never heard of such a thing happening in real life. On tends to get the impression, that it is a standard phrase used by newspaper reporters, when they can't be bothered to do any actual research.

Anyway, things kind of peaked on the 9th of July, when a Swedish couple claimed to have filmed the lion. Experts were not convinced and said it looked more like a lynx than a lion. They were equally unconvinced by some of the tracks people claimed to have found. Some of them looked suspiciously like they had been made by pressing a spoon into the ground.

Whatever had happened, the sightings petered out quite quickly, and Elvi withdrew back into the shadows where she had come from.

Sightings and guns
In Finland the sightings tend to be very concentrated timewise – it is very rare to see

sightings continue for several years in one particular area. This could partly be because the actual number of sightings are in fact quite small, and most of the eye-witnesses have in fact not seen anything special, but have just been carried away with the general excitement. Or, which I am inclined to favour, the explanation could be, that these cats simply cannot survive a Finnish winter.

Several authors have suggested, that sightings of mysterious big cats are either sightings of unknown species, or surviving prehistoric species, or that the sightings have some sort of supernatural aspect. My personal opinion of this entire "circus" is that each and every one of the actual sightings are of escapees, animal that have run away from legal or illegal collections. There may be a small number of hoaxes involved, or indeed misidentifications of other things, but in my opinion it doesn't retract from the general conclusion.

In 1997 there was even a small "flap" of sightings of what must have been an American puma near Pori in western Finland, close to the coast of the Gulf of Bothnia. I have spoken to three people who saw this animal, and all of them stated clearly, that the animal was wearing some kind of collar – it can't get much clearer than that.

Finally there is one other aspect to take under consideration when talking about big cats in Finland. You seldom hear about them outside of the country, and as I have stated above, the various flaps rarely last for long. And this may simply be a result of Finland's rather shall we say relaxed gun laws. The number of guns per inhabitant is far larger in that country than in other parts of Scandinavia. I think most of these big cats simply end up being shot, and I have indeed met several people who hinted, that they had either shot some of these animals themselves or knew someone who had. In all fairness, I have never been able to substantiate any of these stories, so they might be nothing more than urban legends, but still...

The really weird ones
In all collections of sightings, be they of the Loch Ness monster, ufos, ghosts, Men in Black or what have you, there is always a little residue left, when you have sorted, explained and discussed, a little handful of sightings so strange that it is quite difficult, not to say impossible to know what to do with them. And so it is for the Finnish big cats as well. I have two rather strange sightings lurking at the bottom of my files as well. I have tried to ignore them for several years, but they keep bobbing up to the surface when I least expect it and demand attention, so without any further ado, I will bring them here. I think I have an explanation for one of them, but the other one... As always, any suggestions would be most welcome. So here goes...

The multicat
Early one morning in the summer of 1997, a 17-year old girl called Lena, was walking her dog at the banks of Lake Näsiselkä in the eastern outskirts of Tampere in South Finland. She had been out for about half an hour when her dog suddenly started growling, and at first refused to walk another step, but after a few moments started to strain at its leash as if possessed. She had an awful lot of trouble keeping the dog under control although it was in fact only a small terrier. At first she couldn't see anything that could justify the ferocious anger of her normally quite placid dog, but suddenly she heard a strange growling, almost grunting, noise, and smelt an even stranger smell, something that she described as a mixture between an old billygoat and a burning tire.

> *"And then the weirdest cat I have ever seen came out from among the trees on one side of the path, walked calmly across and disappeared on the other side."*

I can easily understand her confusion, because if her description is true, she saw an animal which was a strange mixture of just about all the cats you could possibly wish for – like a cat made by a committee.

First of all it was big at least the size of a female lion, but it had a very strange colour. The head and front parts were very dark brown, almost black, but the colour got gradually lighter along the rest of the body and all the way to the tip of the very long tail, which was pure white. Now lions are normally very powerful animals with fairly short and muscular legs, but this animal had very long, thin and spindly legs, somewhere along the lines of a cheetah or perhaps a maned wolf (which is not a cat, I know, but it has very long legs). There were a few indistinct spots and rosettes on the hindquarters as well. At the back of each leg was a tuft of fairly long black hairs. The head was strangely elongated for a cat, and the ears were very large and upright with black hairtufts at the tips, that were so long they tipped over and hanged down in front of the ears.

When I interviewed the eyewitness in 2010 – she now lives in Malmö in southern Sweden, works in the Swedish police, and is a friend of my local hairdresser, who also lives in Malmö – I was very impressed with the amount of detail she could remember, but she was adamant she had not made anything up, and that she had an excellent memory. I am sure she had, but even the best of minds can fool themselves, so…

She also told me, that she did not feel particularly scared, and that the sighting only lasted for something like 15 or 20 seconds before the animal disappeared again. But it had taken her dog most of the day to calm down again. She never told anyone about her sighting – *"They would probably have locked me up, and thrown the key away!"*, but she did go home and write a detailed description of the animal in her diary.

I did not get any sense of Lena lying or telling me a story just to wind me up. It was quite clear that the experience had made a lasting impression on her, but it is equally clear, that no cat has ever looked like that. I am at a loss to even begin to explain what Lena saw that morning if it was a real animal, but on the other hand, when I interviewed her, she seemed a very calm and clearheaded woman, so I don't know what to think. My only suggestion is that she actually did see some form of cat, perhaps a lynx or something similar, and then over the years, her imagination have subconsciously gradually elaborated on what she saw, to create the animal she described to me. It is a fairly well-known phenomenon, which basically means that the eyewitnesses are not lying as such, they are deeply convinced that whatever they have experienced is true, but they have in fact been lying to themselves, doing a self-contained version of Chinese whispers lasting many years. It is a phenomenon every birdwatcher knows. If you see what you think is a rare bird, it is vitally important that you write down a description as quickly as possible, preferably on the spot, as even a small delay makes you invent stuff to fill in the gaps you couldn't see. I think many a rare bird sighting has come about in this way.

The flying cat
The second sighting is also rather strange, but at least I have some idea of what the explanation is. It comes from a Dutch woman called Anke Schauer, who in the summer of 2003 had gone on holiday to Finland with her husband to visit a Finnish friend she had worked with in Holland a couple of years before. One afternoon they were taking a walk along the banks of a lake close to the town of Lahti in Southern Finland, where they had hired a small cottage for a few days. Her husband was an eager angler, and he was keen to try his hand at catching Finnish trout. The cottage was an old wooden building located at the edge of a clearing with its back towards the woods and its front towards the lake. Just before they got to their cottage, the path along the lake curved a little bit in among the trees, so they had to approach the cottage from the back. As they were passing the back of the house, something rather strange happened:

> *"A very tiny greyisk brown cat with a thick bushy tail, small ears and big shiny eyes, jumped out of a tall tree a few metres from the house, landed on the ground, ran over to the back wall, and started walking up it like it had never heard of gravity. We were both looking at it with open mouths. We had never seen anything like it before. None of us are especially interested in animals apart from my husband's fishing and the*

fact that we have kept a couple of cats. But I am certain it was some kind of cat. The cat kept on going until it reached the roof – and then it disappeared, We ran around to the other side of the house where a ladder was leaning on the wall, and my husband almost ran up it – but although it could only have taken perhaps 20 seconds, the cat had completely disappeared. To this day I don't know what it is, and we never saw it again. We tried to telle the story to our friends, but they had no idea what it could have been."

Well, in this case, I might have a suggestion as to what that animal actually was.

In general the Finnish fauna is not that different from the rest of Scandinavia, although there are a few interesting exceptions, one of which is the flying squirrel, which is fairly common in some parts of the country, especially the south. Flying squirrels naturally spend most of their time in the trees, but do occasionally come down to the ground, where they can move surprisingly quickly. I have seen them on a couple of occasions, and they look very much like described by the eyewitnesses. As they are mainly nocturnal, they have rather large eyes – and small ears like most squirrels, and they would have absolutely no trouble walking, or rather climbing, up the side of a wooden building. The only thing that surprises me is the fact that the eyewitnesses did not see, or at least pay any attention to the flap of skin the squirrels uses when "flying" or rather gliding. It usually bunches up along the sides of the animal when it is running or climbing and making it look like it is wearing an ill-fitting jacket a couple of sizes too large.

It is rather a pity the eyewitnesses did not get a closer look at the animal, as flying squirrels are very charming animals that can get very tame. I have been lucky enough to be the guest of a family who had found an injured flying squirrel and nursed it back to health. When it was well again, it absolutely refused to go back to the forest, so they kept it in their house, where it had a free run of their bookshelves, potted plants and ceiling lamps, from where it liked to launch surprise attacks on unsuspecting guests, usually landing on their shoulders, but occasionally on top of their heads – (it had very sharp claws which it digged into my scalp, when I gave a start, when it landed). From here it would descend to whatever pockets your clothing had and search them thoroughly for any kinds of treats. In my case it found a half-eaten package of mints, proceeded to dismantle it, and tried to eat one of the mints. That unfortunately did not suit its refined tastebuds, which prompted the squirrel to deposit the mint down the front of my t-shirt. Good fun, but slightly lacking in general manners I would think.

And finally – a trio of... who knows?

I have spent a lot of time and space in this chapter on various creatures where there has been a reasonable number of sightings to work with, but as a sort of final antidote, I would like to present three creatures of which I only have very scant information, but all of them sounds like they are definitely worth a closer look.

The father of all tortoises

In his now classical work, *"In the wake of the sea-serpent"* Bernard Heuvelmans describes a type of sea-monster that looks very much like a giant turtle. It is very much an animal of the sea, and as such have no place in this book, but according to an elderly Finnish gentleman I met in the Swedish town of Sävsjö in 1989, the Heuvelmans animal has a land-living colleague in the remote north-eastern part of Finland. According to my informant, this huge animal, about as big as a good-sized truck, is well-known to the Sami people, although they prefer not to talk about it, and the Russians on the other side of the border is supposedly well aware of its existence too. The animal is very real, but also mythological, in being a sort of father of all turtle and tortoise kind. I was naturally extremely interested in whether my newfound friend had actually seen the thing, or if he knew someone who had, but when I pressed him for information, he seemed strangely reluctant to discuss it any further, and instead did his best to persuade me to try some of his homemade garlic vodka, which was absolutely vile, but I gamely had a couple to try and get him talking. Finally he relented and explained, that the reason he wouldn't talk about the animal was, that it would bring him bad luck – and especially if he told me its name. The father of all tortoises was just a translation of its real name.

I have never met this distinguished gentleman again – mainly because he died two years later, possible cursed by the giant tortoise, but more likely because he was in fact 94 years old. One of my friends from university had a holiday cottage nearby, and kept me informed of what went on in Sävsjö. But I have talked to a good number of Finnish people since then, trying to get more information, but I have never met another one who had heard anything about the father of all tortoises, so I still don't know whether he was just winding me up, or whether there was actually any substance to the story. Needless to say, the Finnish/Russian winter would kill any turtle or tortoise, so it is a bit strange that the story should be about an animal like that.

The king marten

In the winter of 1959, A.T. who at that time was 16 years old, borrowed his father's small rifle and went hunting in the forest west of Oulu where he lived at the time. He hadn't any real plans, or as he later admitted, *"any real idea of what I was actually doing. I think I had this vision of returning with some kind of fantastic bounty, which would me a hero, make everybody like me, and a girl I fancied at school fall madly in love with me!"* Being it the middle of winter the amount of light was rather small, so he managed to get lost very quickly. But determined to prove himself, he plodded stubbornly on, and was suddenly rewarded with a strange growling sound that made the hairs at the back of his neck stand upright.

"I had never been afraid in the forests before that day. I don't know why, but the sound of that growling almost made me wet myself, and I froze completely in my tracks. At first I couldn't see anything, but then I suddenly realized that about 5 metres straight in front of me an animal was looking at me. It could only see its head, and I was in no doubt I was looking at a marten, except for the fact that the head was the size of the head of a big dog. I had the distinct feeling the animal was about to attack me, so I lifted the rifle and fired without thinking. I hid the thing directly in the head and it dropped down like a sack of potatoes. For the next few minutes I was completely unable to move. My heart was pounding so hard and so fast I thought I would faint. When I finally calmed down I walked over to the animal on rather shaking legs and took a

closer look. It was a marten, but it was the biggest I had ever seen. The head was the size of big dog, and the body was very thick and powerful, and close to 3 metres in length. The legs were also rather long. I knew how an otter looked, and thought I had killed one of them, but the legs of this animal was so long it head was more than a metre of the ground. So it clearly couldn't be an otter. I thought my father would know what it was, so I tried to pick it up and carry it, but it was so heavy I could hardly lift it. So I decided to come back the next day and pick it up. I covered it with branches to keep other animals away, and then I started to find my way back. I took several hours, and my father was not pleased when I got home.

The next day I went back, but I couldn't for the life of me find the animal again. Or rather I think I found the place where I had shot it, but there was no trace of it. The branches lay scattered all over the forest floor, and all that was left of the animal was a couple of tufts of dark brown hairs. Maybe a bear came by and ate it, or perhaps wolves, I don't know. I never saw anything like it again. And almost everyone I talked to about it thought I was lying. The few that actually believed me had no idea what the animal could have been. I still don't know what it was. Do you?"

The answer to that question I am afraid, would have to be a resounding: "No!" But if anybody else has, or have heard of a similar animal anywhere else, I would love to hear from them.

And finally – a big bird

I am big fan of the ostrich and its relatives. I have seen them in many parts of the world, and have always enjoyed them immensely, especially close encounters with Australian emus, the only birds that I know off that have to be characterized as completely bonkers, and the cassowaries, the strange black rainforest birds with their enormous bony crests and colourful wattles. Most of these birds live in the tropics, so imagine my surprise, when I heard a story about a family who had seen what they described as an enormous ostrich with the head of a horse.

The story is tantalizingly short, and frustratingly intriguing, but for what it's worth:

In the summer of 1998 a Swedish family on holiday in Finland was driving along the west coast of Finland somewhere between Vaasa and Pori. Everybody was a bit tired and bored, and was looking forward to get to their destination, get a shower and a good meal. They had been driving through a bit of rather dense forest for at bit, but suddenly the trees retreated and they could see an enormous clearing. They could also see an equally enormous bird running across the clearing with large bounding strides. It was quite a long way away, and the car was moving, so they only got a rather fleeting glimpse of the thing, and by the time the father who was driving managed to stop the car, the enormous bird had disappeared. The all described it as looking like an ostrich, but much bigger and much more powerfully built – and with a head the size and shape of a horse's head. The only dissenting voice was the youngest son in the family, a boy of 14 who had good eyes. He told me, that as far as he could see, the "head" was actually an enormous beak.

The family discussed briefly whether to try and perhaps find some tracks of the bird, but it was late in the day and they were all tired, so they decided to move on – something the son told me they have regretted deeply, as they have been at the receiving end of numerous jokes since then.

So what did they see? A fleeting glimpse of a deer that for some reason made them think of a bird? An actual horse? Or where it just a case of mass hallucination brought on by general fatigue? Or did they actually somehow get a glimpse of a supposedly long extinct terror-bird? And if so – how?

A nice, friendly and slightly melancholic afterthought

No chapter on mysterious Finnish creatures, be they real or imaginary, would be complete without a least a fleeting look at what is probably Finland's best known legendary creatures – The Moomin of Mumitrolls, as they are known in Scandinavia. These white, pot-bellied troll-like beings with big noses that make them look like upright hippos, were the creation of Finnish author and illustrator Tove Jansson. The stories of their gentle lives which always makes me think of Winnie the Pooh, were first published in Sweden i 1945, and were for many years perennial bestsellers in Scandinavia and the Baltic countries. From early 1990's and onwards, the interest in the Moomins have become a worldwide phenomenon, and many people have speculated on the origin of the Moomin. Were they in fact, as some have suggested, bases on creatures from Finnish folklore?

I would love to be able to report that some people have claimed to have seen real life Moomins, but alas, I can't. In various interviews Tove Jansson did claim that a substantial part of the characters were in fact inspired by friends and family members, and various writers have suggested that the actual Moomins were in fact based on various aspects of Jansson's own personality – and not mysterious creatures lurking in some remote part of Finland. Sorry!

THE BALTIC COUNTRIES - Three problematic neighbours

The Baltic countries are a bit of a problem in many ways, when it comes to a book like this. (I am in no way talking about anything even remotely political.) It is not because they are lacking in folklore, sightings of strange beings of interesting history – natural or otherwise – far from it, but there is a distinct language barrier between these three countries and the rest of northern Europe – with Finland as a possible exception, but that's only because some part of that country has a language barrier all its own. So since most of the literature on folklore and similar subjects are written in one of other of their native languages, it only has limited accessibility. That is most unfortunate, as I am sure there is a vast amount of exiting stuff waiting to be discovered, if I can in any way extrapolate from the relatively few stories I have managed to dig up via contacts with students and tourists coming to Denmark from either of the three countries, from correspondences and interviews with actual inhabitants, only few of which speak any of the languages I speak. I have already shown in my earlier book "Weird Waters" (CFZ Books 2011) that there is a fair number of sightings of unusual aquatic creatures, but the number of sightings of terrestrial things are apparently considerably smaller. The amount of material is actually so small, that it would be meaningless to give each country a separate chapter, so in this book, they will be treated together in one chapter.

The sightings that I have been able to collect, although few in number, shows a surprising degree of variation with creatures stretching from full sized dragons to extremely small mice, via some possible prehistoric survivors.

I can only hope, as tourism and travel to the Baltic countries develops even more, that more people with an interest in cryptozoology will go there, and ask about strange creatures, and start to tap the no doubt rich veins of stories that are there. Or that some Baltic author decides to write a book about the weirdness of it all – this would surely bring out more stories. At a couple of occasions people have indicated to me, that there are in fact a lot of stories that are not for foreigner to hear. I wonder what that is all about, but I really hope they will be told at some future date. Until then people will have to make do with my humble contributions to the subject matter.

A short geographical note
Just for future reference – the Baltic countries consists of (going from north to south), Estonia, Latvia and Lithuania, all three countries bordering Russia to the east, and the Gulf of Botnia to the west. Three countries rich in history, but also with a rather tumultuous past, having been attacked and conquered by Viking raiders, and in various points in their history been independent and part of the former Soviet Union.

Here be dragons!!!
And boy do they have dragons! The giant flying reptiles are as a matter of fact so common in the folklore of the Baltic countries, and especially Latvia, that you wouldn't believe it. There are apparently sunken castles, buried castles, disappeared castles, cursed castles and ruined castles everywhere, and each and every one of them is guarded by at least one dragon – in some cases flocks of them. You can also find dragons in deep forests, on mountaintops, inside mountains, in caves, old mines, gravel pits and living in the sky. So – are you into draconology, I can strongly recommend starting to learn the three Baltic languages. I can almost guarantee that you will find material enough to keep you busy for the rest of your life.

The large dragons usually go under the name of pukis – which is a bit strange, as this is also the name of a race of strange goblin-like beings that roam the Baltic countryside. They are very similar to the various little people you find in the rest of northern Europe, and so has to some extent already been treated in this book. I have not been able to find out what connection they have to dragons, if indeed any. But as both types of creatures are very magical, they may to some extent be manifestations of the magic of the Baltic landscapes.

Baltic dragons are classic dragons as it were. They are extremely large and fierce – some of them mindblowingly large – I have heard stories of dragons so big that they actually use the various sunken and otherwise indisposed castles as pillows, nests or perching posts. They are very powerful animals, reptilian, with a long tail, a long neck and a big head with a set of serious carving knives for teeth, although not especially keen on fire-breathing funnily enough. They have four short but powerful legs, with long claws and a huge set of wings, which to some extent wreaks havoc with the idea of dragons being real animals, or at least based on sightings of real animals. If they are indeed reptilian, they are tetrapods, and as such should only have four limbs. If the wings of the dragons are built on the same system as say bat-wings, and according to the various descriptions, that is exactly what they are, we end of with a hexapod, and that is quite a long way out of the realm of scientific possibility – but on the other hand – with a little bit of magic involved, who knows. Come to think of it, the smaller dragontype, the wyvern, fits considerably better into the tetrapod mould, as they only have one set of legs to go with their wings.

The amount of stories and legends about dragons are probably vast, but the few examples I have read or been told, are very much along the lines of what we know from medieval legends in the rest of Europe – i.e. damsels in distress, gallant knights, dragons terrorising the countryside, gathering treasure and so on, and so forth. And I see no reason the repeat that here – that has already been done in much more detail in Karl Shuker's and Richard Freeman's various books on dragons. What to me is a subject of considerably more interest within Baltic draconology, are the puks. These are also dragons, but of a rather different type and temperament than the big marauding monsters we are so used to dealing with.

The puks
A puk is a small and rather benign dragon, mostly known, and rather liked for its predilection to protect specific houses – not castles, not giant fortresses or anything like that, but single, specific houses. The logical consequence of that would of course be a rather large population of puk-dragons, and so apparently it is, or rather was. In olden days, almost every dwelling would have its own puk-dragon as protector. It is easy to see the similarities with house-spirits and other forms of supernatural guardians, but according to some of the stories I have been told about the puks, they were quite straightforward down to earth animals that basically served as flying magical guard-dogs.

Some of the puks were not much bigger than a big dog, but some could be the size of a large horse, but that was it. But one should not let oneself be fooled by their diminutive stature. According to an elderly couple from Riga, the capital of Latvia, I met on the Swedish island of Gotland in the summer of 1990, the puk-dragons were immensely strong, completely fearless, and could not be killed by any conventional means. Nobody quite knew how a puk-dragon selected the house and family it wanted to protect, but one day it would just be there, sleeping on the roof or in the yard, and woe betide anyone who then tried to attack. Having a guardian puk would on the face of it sound like quite a nice thing. In troubled times you would feel protected, but apparently the puks sometimes got a bit too eager in the performing of their duties. Families with beautiful daughters would sometimes have serious trouble when trying

to find a suitable husband for their girls. They and the girl could be as fond as they would of a suitable party. If the puk didn't like the boy, the wedding was off.

The stories about the puk-dragons are old quite old, dating at least back to Medieval times, and possibly even longer, and I have never heard of anyone claiming to have actually seen one. But nevertheless, even today dragons are very much part of the scenery. Go for a walk in f.inst. Riga, and you will meet dragons sculptures, weathervanes and paintings all over the place. Even St. George and the Dragon are depicted in various versions in the city. And if you get to Tallinn, take a look in the Glehn park – there is a allmighty great stonedragon on the rampage, and it's a big one!

The furry ones

You don't have to go many decades back in time before you get to a period when wolves were common all over the Baltic countries. With that in mind, it is perhaps not surprising that one particular monster, or mysterious creature if you like, is very common in the local folklore, and that is – I am almost tempted to say of course – the werewolf. From some stories you get the distinct impression that there was a time, when half of the wolves were in fact werevolwes, and half the human population were suspiciously hairy, and with a strange predilection for nightly perambulations when a full moon was approaching.

A werewolf is generally considered to be a human cursed by some form of dark magic and forced at every full moon to turn into a ferocious werewolf, so much in the grip of its own insatiable bloodthirst, that it is capable of killing everyone it meets, even its own loved ones. The transformation was also supposed to be extremely painful, which would go some way to explain the rather foul mood of your average werewolf, so in general not something to be wished for in any way.

Although extremely dangerous, werewolves were in general creatures – or humans if you like – to be pitied in spite of the atrocities some of them had committed. And one could almost see it as a sacred duty to put them out of their misery, with whatever means were at hand, the most efficient generally thought to be shooting them with silver bullets, preferably made from consecrated church silver – talk about recycling!

There was also another kind of werewolf, thankfully an extremely rare one. In these cases the human had actively embraced the dark magic voluntarily, and were capable of transforming themselves at will. This sounds sinister, but not all of them were in it for nefarious purposes. One alleged werewolf, who were caught in Latvia in 1697, was adamant that he had never killed anyone but hares and gamebirds, when he had transformed, and that he only turned himself into a wolf because it gave him an immense sense of freedom to be able to run like the wind, possibly augmented by the fact, that when in his human form, he had a twisted leg and a severe limp – oh yes, and because the "sexual congress with a female wolf" was apparently out of this world!

By and large though, the Baltic werewolves seemed to have been slightly more peaceful, or at least slightly less bloodthirsty than their European counterparts. I have searched in vain for stories of killing sprees along the lines of the infamous French case of the Beast of Gevaudan (1764-1767) where something like a wolf, a dog or a dog-wolf hybrid or indeed a werewolf, took to killing French peasants with gay abandon. One does find stories about people being found horribly mutilated by some ferocious predator, but only occasionally, and nobody seems to have seen any of the actual attacks taking place. And the people accused of being werewolves were probably mostly harmless eccentrics or people with various psychological problems.

I have been told quite a few stories about sightings of wolves rising up on two legs and starting to walk like a human being, or indeed humans going down on all fours and preceding to outrun even the fastest hunting dogs. I don't quite know what to make of these stories, partly because I have a sneaking suspicion, that they are in fact different variations of the same story, some sort of Baltic urban legend if you like.

They (it) are all from the early 1700's, all from the eastern part of Lithuania, and they were all made by 16 or 17 year old girls. One cannot help but wonder whether these sightings tell more about the state of the girl(s) than what was actually going on at the time.

I have no doubt that people were in fact sometimes killed in rather gruesome ways, and there could have been many reasons for that – rabid dogs springs quite easily to mind – but I am also quite convinced that wolves in most cases simply were the closest and most suitable scapegoats. If ever there has been an animal in need of a PR-manager and a couple of spindoctors, it is indeed the much maligned wolf.

An intriguing possibility, suggested by several authors, present company included, is that werewolf stories to some extent have their basis in garbled stories about Viking raiders who would sometimes dress themselves in wolf-skins. I can quite easily see a berserker warrior with the skin of a wolf attached to his helmet, become a bloodthirsty half human-half wolf a few generations down the line.

The king wolf
Weirdly enough, and now, with your permission, we take a small and hesitant step into the realm of cryptozoology, there may in fact be a somewhat more factual zoological explanation for the werewolves, or at least the very gruesome attacks on humans – and in some cases livestock as well. There are stories about fully grown bulls being almost bitten in half and similar carnivorous mayhem. I base this assumption on the fact that I have four stories about what people have called king wolves on file, and even in the otherwise spooky world of wolves and wolflore, they are a bit unusual. Two of the sightings are from Lithuania, and one each from Latvia and Estonia. Three of the sightings are from the middle of the 1800's, but the one from Estonia is from 1972, and thus worth bringing in full I should think. The other three are quite similar, although not as detailed.

The witness is a venerable schoolteacher, who today teaches mathematics at a school in Tallinn, the capital of Estonia, but in 1972 was a 16 year old boy living just outside Vänska in eastern Estonia close what is now the border with Russia.

"It was late January, in the middle of winter. It was absolutely freezing cold, and my mother had told me to go out and collect a sackfull of pine cones. They were good kindling, and we need all the heat our fireplace could muster. So I grabbed a sack and headed for the woods nearby. I had to walk a fair bit, as I was not the only one who had been out to collect cones. It was a very clear day. The Sun was out, and despite the cold some birds were out and about trying to find some

food. I could hear a lot of chirping and tweeting, and one of two crows or ravens croaking away somewhere. After having walked for about half an hour, I found a good spot at the edge of the forest, close to a large section of open ground with a low ridge running through it. There were lots of cones on the ground, so I started picking them up and putting them in my sack. I had only been doing it for a couple of minutes, when I realized, that the whole world had suddenly gone quiet. There was not the slightest sound anywhere. But just as suddenly, an almighty racket started up. It sounded like 200 ravens and crows had starting a croaking contest. I started looking around to find the cause of all this noise, and saw a huge flock of black screeching birds suddenly rising up from the ground, followed a few seconds later by the biggest wolf I have ever seen. I swear to you, it was bigger than a bear. This had to be one of those king wolves my grandmother had told me about. All I could do was stand there and stare while this huge animal turned and walked along the ridge until it disappeared among the trees. It looked just like an ordinary wolf, only bigger. I think I could have walked under its belly without bending down."

After this rather extraordinary experience, the boy hurried home, rather shaken, and did his best to forget all about it.

So – a giant wolf? Perhaps made bigger by equal amounts of fright and a bit of optical illusions – I have often experienced the strange sensation of things looking much bigger than they are in the cold, clear air of the Arctic, or indeed in the winter when you go north in Scandinavia. Was it something like that my contact had experienced – perhaps augmented by the fact that this particular wolf was indeed very big? I don't know, but every now and then most animals turn out an extremely large individual, so could such a phenomenon be the explanation both for the king wolves and the rarity of them? Whatever the explanation, there is no doubt that a very large wolf would be capable of the killings ascribed to werewolves.

Following the rather scary story of the king wolves, I thought we could go on to something even scarier. It is time to meet the ööitketaya – the nightwalkers.

Be afraid, be very, very afraid!
Have you ever woken with a start in the middle of the night, scared out of your wits without knowing why? Maybe not as an adult, bus as a child I'll bet you have, and more than once. You may even have felt sick and miserable without any specific reason. Well – now the truth can be revealed. You have been attacked by the

ööitketaya – or the nightwalkers, in case you don't speak the local language.

The nightwalkers are a whole race of evil formless spirits consisting of nothing but blackness and malevolence. They roam at night looking for people to torment, and they are especially fond of attacking children, but are quite capable of scaring the life out of adults as well. They have a distinct similarity to similar terrifying nights creatures from other parts of the world who make a living out of scaring people or giving them nightmares or a general nocturnal hard time. The main difference that sets the nightwalkers apart from similar creatures is that you can actually every now and then be unlucky enough to meet them in person or spirit, roaming around at night – not just whilst laying in your bed.

I only know of two actual sightings that wasn't made by people just waking up and feeling or seeing dark figures in their room – monsters under the bed? That's nightwalkers too. I am sure some researchers would see them as evidence of alien abductions going way back in history, but I see them as nothing more than lingering vivid or lucid dreams.

Vidid dreaming is one thing, but actually seeing the nightwalkers in the flesh or whatever they are made of, is quite another matter. Now as far as the two sightings I have in my files, I would have no hesitation in calling them something inspired by the dementors in the Harry Potter novels by J.K.Rowling, were it not for the fact, that at the times the novels were still many years in the future.

A nightmare of twins

For the actual sightings we have to go back to 1973, to a family living in Vääna some 15 km southwest of Tallinn. For several nights in late summer, a couple of 7-year old twin boys had been having violent nightmares that had them waking up screaming and absolutely refusing to go to sleep again, claiming something evil was looking through their window, making their room grow cold, and making strange wheezing noises. In an attempt to calm them down, their two older brothers aged 16 and 17 offered to patrol the street outside their house and the window to the boys' bedroom.

As it was a rather warm night in august, the two older brothers were not too bordered by their vigil, but was standing in the street chatting, not expecting anything much to happen. But something did in fact happen. I have heard the testimony of both brothers, and they are very alike, so for the sake of brevity I will only bring one of them, 18 year old Ennis:

> *"We were standing outside the house, when our brothers suddenly started screaming hysterically. At first we couldn't see anything strange at all. A tree or something was casting a shadow on the wall of the house and across the window to the bedroom, and at first we thought that was what had scared our brothers. But then I realized that there was nothing near the house to cast that shadow. I kept looking for a tree or a post or something, but there was nothing, nothing! We were both staring like mad at the house, when the shadow suddenly let go of the house, almost like something was sucking it away. And then, I am sure I imagined it, but it looked like the shadow took two steps back from the house, and then disappeared or dissolved itself. We looked at each other and decided not to tell our parents that we had seen a nightwalker. They would never believe us. But somehow it helped. From that night on, our brothers could sleep again. Maybe by seeing it, we scared the nightwalker away. I don't know. But I know that for the rest of that summer, I only when out after dark when I was with my brother or with someone else – never alone."*

That was probably a wise precaution, because according to the stories about the nightwalkers, they can also kidnap people. Unfortunately, unlike other kidnappers, the nightwalkers never ask for a ransom, and they never let their victims go.

I think the nightwalkers are some form of, shall we say personification of the dark and the natural fear of the dark, an attempt to form some kind of explanation to night terrors and people who disappear without a trace. Bogeymen if you like. The only trouble is that bogeymen as a rule do not exists, so what did the two older brothers in fact see that night? I don't actually think they saw anything. They knew the stories of the nightwalkers, and had in facts discussed them earlier on, so I think they somehow led each other to believe, that they saw a nightwalker leaning on the house, scaring their brothers. They insisted there was nothing close to the house capable of casting a shadow, but on the other hand they also said it was a fairly clear night with a bright, although not full moon, so perhaps a shadow of a small low cloud combined with a bit of exited imagination?

Anyway – enough of the scary stuff and time to get on to some proper cryptozoology. Because my search for interesting stories have also turned up some sightings of a couple of rather well known prehistoric animals nothing less than the aurochs and the giant elk. Strangely enough, I had never heard about them in a cryptozoological context before I started writing this book. But lo and behold, a little handful of people actually thinks they have seen them several hundred or thousand years after they officially disappeared.

A mighty ox
The last officially known aurochs died in 1627 in a forest in southern Poland. In one sense, it is just another sad loss in the endless row of animals that humans have exterminated, but it has a special significance, as the aurochs was the ancestor of almost all of today's domestic cattle. On a cryptozoological level, the aurochs is interesting, because there was always the off chance that some individuals had actually survived for a longer period in others areas of its distribution. You don't normally hear about aurochs sightings, but a few did actually surface when I started digging – and all of them concentrated in one particular area, what is now the Dzukijos National Park in Lithuania close to the borders with Belarus. One group of sightings is from 1777, where during one summer, several people allegedly saw two very large almost completely black bulls at various places in the forests. The animals were extremely shy, so they couldn't have been ordinary escapee bulls from some farm. Local people were certain it was aurochs. The animal was after all at one time considerably more commons and could be found in a large area from the big forests of Central Europe and further east into Russia. Unfortunately there is not much in the way of details when it comes to these sightings, but I see no reason why a small population should not have survived for a century or so longer. And if the final sighting is anything to go by, they did in fact survive for something like 250 years beyond their official extinction date.

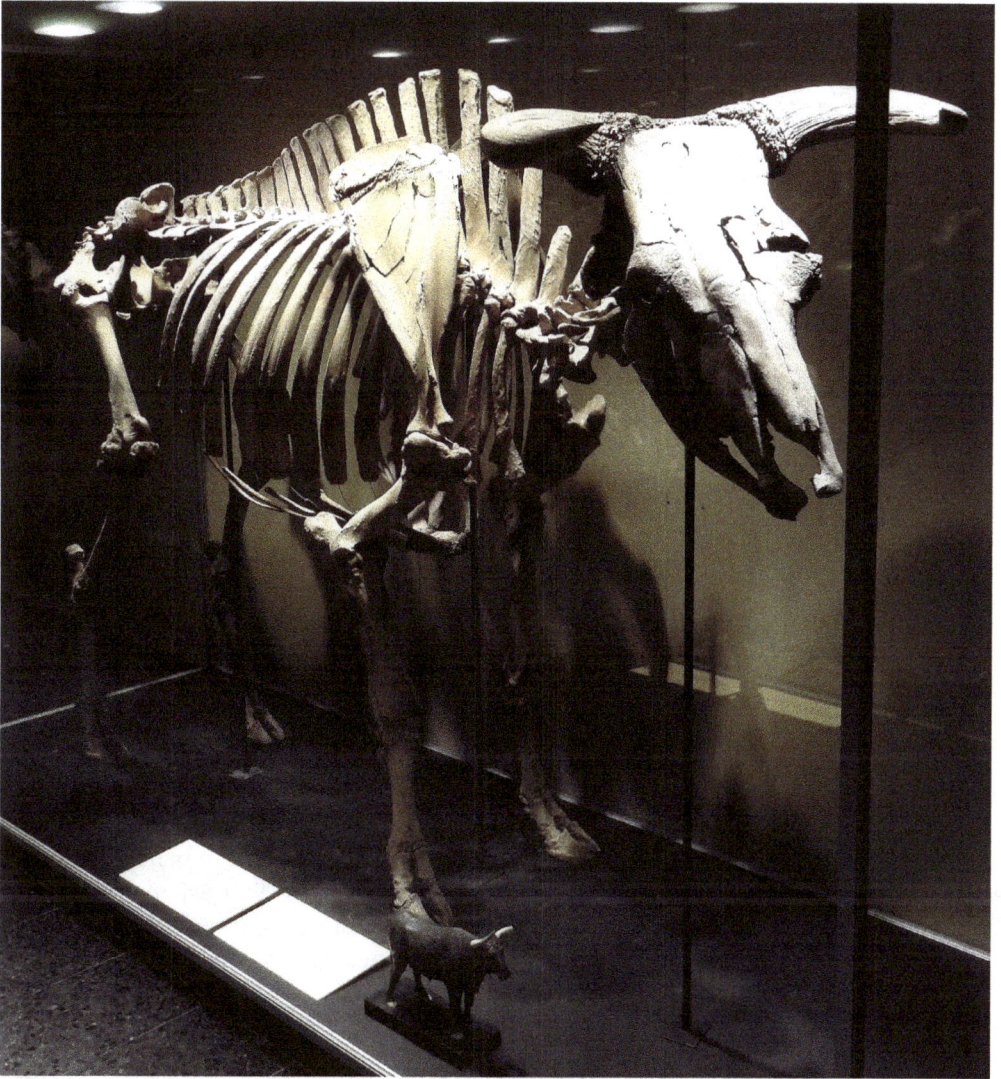

In December 1962 a local man out hunting for birds in the Dzukijos Forest found a series of very large cattle tracks in the snow in a clearing in the forest. He started following them, thinking that one of his neighbour's cows had escaped. The animal had wandered in an irregular fashion in and out among the trees, and did not seem to be going in any particular direction. It took him close to an hour to catch up with it, but when he finally did, he was in for quite a surprise.

Excerpt from a translated transcription of a diary, given to me by the hunter's daughter in 1998:

> *"The cow was very large and as dark as Tiko (the family's black dog). It was tall like me, and had very long legs. It just stood there and looked at me. It was not afraid. But I was afraid. The horns were very long and very sharp. I thought about shooting it, but my birdgun was too small. It would only tickle it, I think. And then maybe it would attack me. Tiko did not like the cow. I think he was afraid also. In the end I just decided to go home. This could not end well. Perhaps I can try to find it some other day and bring a bigger gun."*

I don't actually know if he tried to find the big cow again, but according to the daughter he never talked about shooting it, so she presumed he never did.

A stag the size of a mountain

I have already touched on the possible continued existence of the giant elk with a sighting described in the chapter about Greenland. I find it extremely unlikely that something that big could hide out for so long, especially since the giant elk must have been an animal of open country considering the enormous size of the antlers it carried around. So even if it did hang on in remote areas of Siberia or similar places for longer than what is normally considered the extinction date (about 7700 years ago) there would surely have been a fair number of sightings by now. And in fact, apart from the sighting from Northwest Greenland, I know of only one more – from the extreme east of Estonia in 1956 – as told to me by the man's daughter in 1999, when she was 59, and her father had been dead some 20 years.

"My father had gone out to collect mushrooms in the forest. He was very good at it, and he knew a lot of places where there was plenty of mushrooms. For as long as I could remember he had done so without any problems, but that day he came home looking like a ghost. He

234

was pale and shaking, and my mother had to give him a glass of vodka before he could talk. He told us, that he had gone to a place at the edge of a forest, where he usually could find lots of good mushrooms. He had found some, and put them in his basket, but then he suddenly heard a strange braying sound. He had gone behind some trees at that moment, and through the branches he could see a very large deer running across the open land. He kept saying it was big as a mountain, it was big as a mountain. He told my mother and my sister and me, that the deer had the biggest antlers he had ever seen. He said they were so big a grown man could be carried in them. The animal was bigger than the biggest moose, but it did not have the long nose of a moose. It looked like a normal deer, but very, very big."

I have no idea as to what the man actually saw. The most likely explanation is of course a very large red deer stag, but even though big stags can have a very impressive set of antlers, they are hardly big enough to enable a grown man to stretch out in them, whereas the giant elk did have enormous antlers. He could of course have misjudged the size of the animal to a fantastic degree, but he was used to being outside, going hunting, gathering mushrooms and so on, and his daughter was quite adamant, that he was terrified when he came home. As far as the daughter knew, her father was the only one to see the giant deer. They never talked about to their friends and neighbours for fear of ridicule, but she never forgot that story, and neither did her father. He returned to it regularly for the rest of his life.

The minuscule minute mice

One of the strangest stories from the Baltic countries I have been able to lay my hands on, comes courtesy of a birdwatching friend of mine, who spent a whole month birdwatching in the area in the summer of 2010. At one stage during this time, he was staying in a small hotel in central Latvia, using it as a base to visit various interessant areas. He had with some difficulty managed to explain to his hosts, and elderly couple in their 70's, what he was doing, and that inspired them to regale him with stories about birds and other animals, all of them rather strange. Among many other things they insisted, that birds did not fly south in the winter, but spent the time sleeping in hollow trees, in caves or even underground. They even claimed to have found a whole flock of crows sleeping in their attic in the middle of winter some years ago. They also told him, that female martens did not have to mate, to get pregnant, all they had to do was to stand downwind of a male, and that would be that.

They also had a good method for my friend to become rich. They were well content, so they did not need help, but if he were interested, he should seek out a special type

of mice to be found in clearings in the forest. These mice were tiny – and I do mean tiny. According to the old couple, they were only 1 og 2 centimetres in length, they lived in small groups in the ground, and you only saw them in the very early spring, where they would crawl around on the ground.

This was the time, where you had to catch them, and gently stroke them with a small brush, because the miniature mice would have a sprinkling of gold dust in their fur, and if you were patient enough, you should be able to gather a decent amount of gold.

I am reminded of the Greek historian Herodotus' tale of big furry ants living somewhere in Persia with the ability to dig out gold-nuggets from the ground. It is a classic in folklore, and the gold-digging ants can be found in many a Medieval bestiary. But gold-digging mice? That was a new one.

I suggested to my friend, that the "mice" the old couple were talking about, were in fact bumblebees, and their way of crawling around very early in spring is quite typical. In spring some days it is simply too cold for them to fly, so instead they crawl around from one springflower to the next, trying to gather the first nectar and pollen of the year. Now bumblebees are hairy, and pollen is often yellow or yellowish, or dare I say it – gold-coloured, so I think I am not too far off the mark with the bumblebees.

Thus endeth the lesson
On this rather weird note, we have come to the end of our journey through the wilds and wonders of Scandinavia and the neighbouring areas in Northern Europe. I hope I have demonstrated, that this is a place heaving with interesting stories and weird creatures, and also a place with plenty of possibility for further study. I have only scratched the surface, but have hopefully exposed enough "ore" to make others follow and find even better tales and legends – and perhaps even actual physical traces of some of the creatures. It might just be a weird insect or a strange bird, but just imagine what it would be like to stare a hiisi in the eye or perhaps catch a nightwalker of a dragon if such a feat is even possible. Who knows? Have fun!

SOURCES:

Books, magazines and newspapers:

- Agergaard, Erling: Bogen om trolde. BoD 2009
- A to Z World Superstitions and Folklore, World Trade Press
- Anker, Jean & Svend Dahl: Fabeldyr og andre fabelvæsener I fortid og nutid, Povl Branner 1938
- Benedikt, B.S.: Basic Themes in Icelandic Folklore, Folklore 84 (1), spring 1973, 1-26
- Bergsøe, Wilhelm: Fra Mark og Skov, Gyldendal 1880
- Boucher, A. ed.: Elves, Trolls and Elementals Beings, Iceandic Folktales II (1977)
- Bringsværd, Tor Åge: Phantoms and Fairies from Norwegian Folklore, 1970
- Det "moderne" Urmenneske, Berlingske Tidende June 1929
- Downes, Jonathan: The owlman and others, CFZ 2006
- Eberhart, G.M.: Mysterious Creatures, A Guide to Cryptozoology, ABC Clio 2002
- Ebbesen, Finn: Nyt vidne har set lossen, Morgenavisen Jyllands-Posten, 30-1-1998
- Eilertsen, Mogens: Monsterbogen, Frydenlund 2012
- Elf guide to Reykjavik, Fortean Times 74: 16, 1994
- Feilberg, H.F., Nissens Historie, Wormianum 1979
- Folklore Fellows' Communications, University of Helsinki, Dept. of Folklore Studies
- Forten Times 67 (17), 1993
- Fortean Times 139 (23), 2000
- Fortean Times 141 (23), 2000
- Fortean Times 281, 2011
- Freeman, Richard: Dragons – More than a myth?, CFZ Books 2005
- Gandrup, Niels: Dræberdyret er en los, Jydske Vestkysten, Esbjerg 3-3-1998
- Hannestad, Steinar: Ørnerovet, Lunde 1960
- Helander-Reinvall, Elina: Sami Mythic Text and Stories, 2004

- Heuvelmans, Bernard: In the wake of the sea-serpent, Rupert Hart-Davis 1968
- Hlíðberg, Jón Baldur & Sigurður Ægisson: Meeting with monsters, JPV útgáfa 2011
- Hofberg, Herman: Svenska folksägnar, NILOE 1983
- Ingulstad, Frid & Svein Solem: Troll. Det norske trollets forskrekkelige liv og historie, 1993
- Kristensen, Evald Tang: Danske Sagn, Nyt Nordisk Forlag 1980
- Kvideland, Reimund: Norske Eventyr, Universitetsforlaget 1972
- Lavers, Chris: The Natural History of Unicorns, Granta 2009
- Lycke, Michael: Fauna Ignorata – Nisser & Trold og andre glemte dyr, Hovedland 2000
- Magnus, Olaus: Historia de gentibus septentionalibus (The story of the Nordic People) 1555; Swedish edition: Historie om de nordiska folken, Michaelisgillet 1909-1925
- Nihlen, John: An Isle of Sagas: Legends and Folklore from Gotland, Gotland Turistförening 1931
- Nordic news, Fortean Times 43: 45, 1985
- Not in the best of elf, Fortean Times 93: 20, 1996
- Pilkington, Brian: Islandske Trolde, Gads Forlag 2000
- Putzi-Ortiz, S.: World Superstitions and Folklore, World Trade Press 2009
- Rasmussen, Knud: Myter og sagn fra Grønland, Sesam 1994
- Roestad, Bente: Ørnerovet på Leka, Damm 2006
- Rosen, Sven: Out of Africa – Are there lions roaming Finland?, Fortean Times 65: 44-45
- Rosen, Sven: Joan Petri Klint, A Swedish proto-fortean of the 16th century, Fortean Studies vol 1. 1994
- Sockerlind, R: Elves in Modern Iceland, www.ismenet.is
- Shuker, Karl: Dr. Shuker's Casebook, CFZ Press 2008
- Shuker, Karl: Dragons – A Natural History, Simon & Schuster 1995
- Stattin, Jochum: Tro och vetanden om skånska väsen, in J.Stattin (ed.) Det farliga livet, Natur och Kultur 1991
- Stefansson, V: Icelandic Beast and Bird Lore, Journal of American Folklore 19, 75, 1906
- Stenquist, C. & M.L. Cronberg (ed): Dygder och laster. Nordic Academic Press, Lund 2010
- Studia Fennica Folkloristica, University of Helsinki, Dept. of Folklore Studies
- Svensson, Richard: The serpents of Sweden. Fortean Times 264
- Tabte hår kan måske afsløre mystisk dræber, Skive Folkeblad, 5-2-1998
- Thomas, Lars: Det Mystiske Danmark (Mysterious Denmark), Aschehoug 2005
- Thomas, Lars: Det Mystiske Danmark 2 (Mysterious Denmark 2), Aschehoug

2007
- Thomas, Lars: På udflugt, Nyt Nordisk Forlag 2006
- Vallance, Jeffrey: Lapp of the Gods, Fortean Times 192: 44-49
- van Grouw, Hein: Some black-and-white facts about the Faroese white-speckled Common Raven *Corvus corax varius.* Bull. B.O.C. 2014, 134(1): 4-13
- West, John F.: Faroese folk-tales & legends. Shetland Publishing Company 1980
- Willumsen et al. (red.), Kystens Fortællinger, Gyldendal 1999

Websites:
http://notendur.hi.is/terry/database/sagnagrunnor.htm
kryptozoologi.no
kryptozoologi.se
www.birdlife.no/fuglekunnskap/pdf/44-havorn-damm.pdf
www.chalquist.com/sami.html
www.folklore.ee/folklore/vol45/gunnell.pdf
www.naturen.no/2010/11/15/rovdyr/Svanhild-hartvigsen/havorn/627406

STILL ON THE TRACK OF UNKNOWN ANIMALS

T he Centre for Fortean Zoology, or CFZ, is a non profit-making organisation founded in 1992 with the aim of being a clearing house for information, and coordinating research into mystery animals around the world.

We also study out of place animals, rare and aberrant animal behaviour, and Zooform Phenomena; little-understood "things" that appear to be animals, but which are in fact nothing of the sort, and not even alive (at least in the way we understand the term).

Not only are we the biggest organisation of our type in the world, but - or so we like to think - we are the best. We are certainly the only truly global cryptozoological research organisation, and we carry out our investigations using a strictly scientific set of guidelines. We are expanding all the time and looking to recruit new members to help us in our research into mysterious animals and strange creatures across the globe.

Why should you join us? Because, if you are genuinely interested in trying to solve the last great mysteries of Mother Nature, there is nobody better than us with whom to do it.

Members get a four-issue subscription to our journal *Animals & Men.* Each issue contains nearly 100 pages packed with news, articles, letters, research papers, field reports, and even a gossip column! The magazine is Royal Octavo in format with a full colour cover. You also have access to one of the world's largest collections of resource material dealing with cryptozoology and allied disciplines, and people from the CFZ membership regularly take part in fieldwork and expeditions around the world.

The CFZ is managed by a three-man board of trustees, with a non-profit making trust registered with HM Government Stamp Office. The board of trustees is supported by a Permanent Directorate of full and part-time staff, and advised by a Consultancy Board of specialists - many of whom are world-renowned experts in their particular field. We have regional representatives across the UK, the USA, and many other parts of the world, and are affiliated with

You'll find that the people at the CFZ are friendly and approachable. We have a thriving forum on the website which is the hub of an ever-growing electronic community. You will soon find your feet. Many members of the CFZ Permanent Directorate started off as ordinary members, and now work full-time chasing monsters around the world.

Write to us, e-mail us, or telephone us. The list of future projects on the website is not exhaustive. If you have a good idea for an investigation, please tell us. We may well be able to help.

We are always looking for volunteers to join us. If you see a project that interests you, do not hesitate to get in touch with us. Under certain circumstances we can help provide funding for your trip. If you look on the future projects section of the website, you can see some of the projects that we have pencilled in for the next few years.

In 2003 and 2004 we sent three-man expeditions to Sumatra looking for Orang-Pendek - a semi-legendary bipedal ape. The same three went to Mongolia in 2005. All three members started off merely subscribers to the CFZ magazine. Next time it could be you!

We have no magic sources of income. All our funds come from donations, membership fees, and sales of our publications and merchandise. We are always looking for corporate sponsorship, and other sources of revenue. If you have any ideas for fund-raising please let us know. However, unlike other cryptozoological organisations in the past, we do not live in an intellectual ivory tower. We are not afraid to get our hands dirty, and furthermore we are not one of those organisations where the membership have to raise money so that a privileged few can go on expensive foreign trips. Our research teams, both in the UK and abroad, consist of a mixture of experienced and inexperienced personnel. We are truly a community, and work on the premise that the benefits of CFZ membership are open to all.

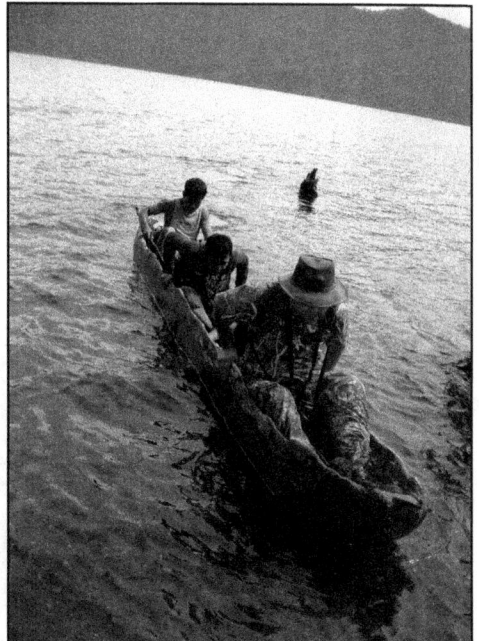

Reports of our investigations are published on our website as soon as they are available. Preliminary reports are posted within days of the project finishing.

Each year we publish a 200 page yearbook

We have a thriving YouTube channel, CFZtv, which has well over two hundred self-made documentaries, lecture appearances, and episodes of our monthly webTV show. We have a daily online magazine, which has over a million hits each year.

Each year since 2000 we have held our annual convention - the Weird Weekend. It is three days of lectures, workshops, and excursions. But most importantly it is a chance for members of the CFZ to meet each other, and to talk with the members of the permanent directorate in a relaxed and informal setting and preferably with a pint of beer in one hand. Since 2006 - the Weird Weekend has been bigger and better and held on the third weekend in August in the idyllic rural location of Woolsery in North Devon.

Since relocating to North Devon in 2005 we have become ever more closely involved with other community organisations, and we hope that this trend will continue. We have also worked closely with Police Forces across the UK as consultants for animal mutilation cases, and we intend to forge closer links with the coastguard and other community services. We want to work closely with those who regularly travel into the Bristol Channel, so that if the recent trend of exotic animal visitors to our coastal waters continues, we can be out there as soon as possible.

Apart from having been the only Fortean Zoological organisation in the world to have consistently published material on all aspects of the subject for over a decade, we have achieved the following concrete results:

• Disproved the myth relating to the headless so-called sea-serpent carcass of Durgan beach in Cornwall 1975
• Disproved the story

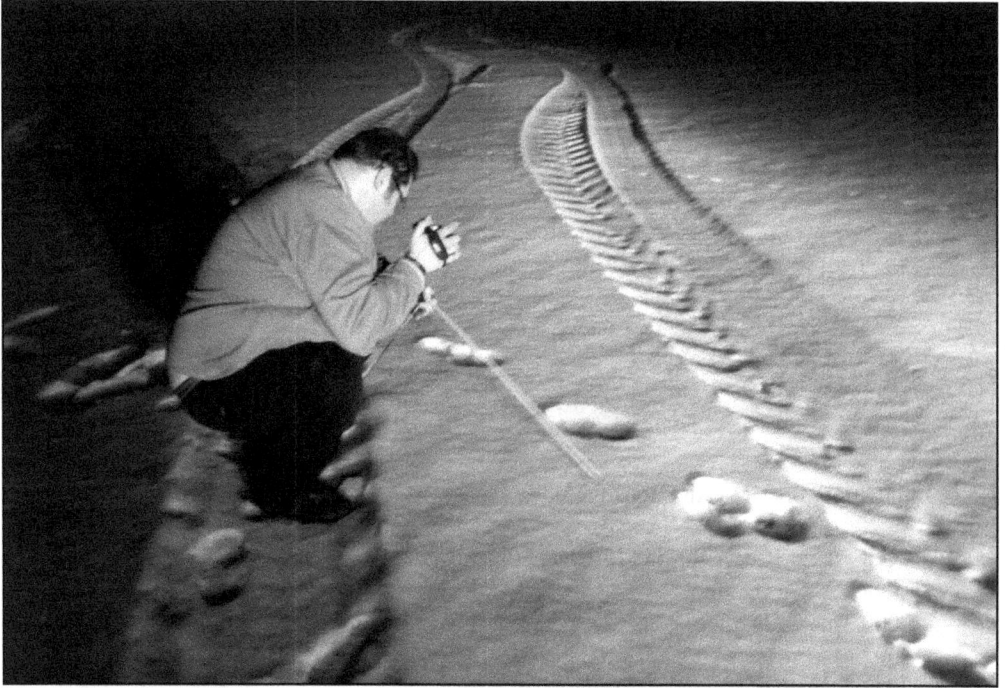

of the 1988 puma skull of Lustleigh Cleave
- Carried out the only in-depth research ever into the mythos of the Cornish Owlman.
- Made the first records of a tropical species of lamprey
- Made the first records of a luminous cave gnat larva in Thailand
- Discovered a possible new species of British mammal - the beech marten
- In 1994-6 carried out the first archival fortean zoological survey of Hong Kong
- In the year 2000, CFZ theories were confirmed when a new species of lizard was added to the British List
- Identified the monster of Martin Mere in Lancashire as a giant wels catfish
- Expanded the known range of Armitage's skink in the Gambia by 80%
- Obtained photographic evidence of the remains of Europe's largest known pike
- Carried out the first ever in-depth study of the ninki-nanka
- Carried out the first attempt to breed Puerto Rican cave snails in captivity
- Were the first European explorers to visit the `lost valley` in Sumatra
- Published the first ever evidence for a new tribe of pygmies in Guyana
- Published the first evidence for a new species of caiman in Guyana

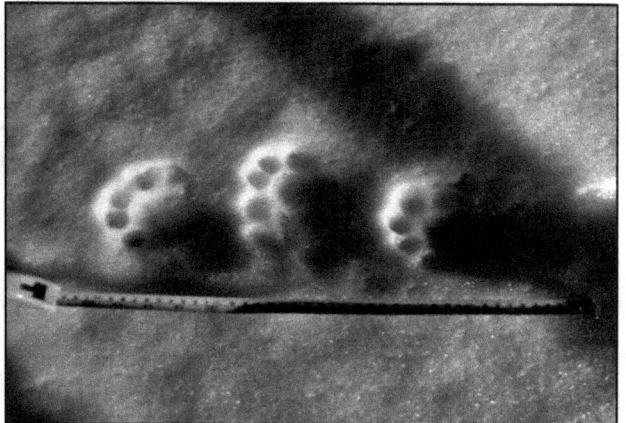

on a monster-haunted lake in Ireland for the first time
- Had a sighting of orang pendek in Sumatra in 2009
- Found leopard hair, subsequently identified by DNA analysis, from rural North Devon in 2010
- Brought back hairs which appear to be from an unknown primate in Sumatra
- Published some of the best evidence ever for the almasty in southern Russia

CFZ Expeditions and Investigations include:

- 1998 Puerto Rico, Florida, Mexico (Chupacabras)
- 1999 Nevada (Bigfoot)
- 2000 Thailand (Naga)
- 2002 Martin Mere (Giant catfish)
- 2002 Cleveland (Wallaby mutilation)
- 2003 Bolam Lake (BHM Reports)

- 2003 Sumatra (Orang Pendek)
- 2003 Texas (Bigfoot; giant snapping turtles)
- 2004 Sumatra (Orang Pendek; cigau, a sabre-toothed cat)
- 2004 Illinois (Black panthers; cicada swarm)
- 2004 Texas (Mystery blue dog)
- Loch Morar (Monster)
- 2004 Puerto Rico (Chupacabras; carnivorous cave snails)
- 2005 Belize (Affiliate expedition for hairy dwarfs)
- 2005 Loch Ness (Monster)
- 2005 Mongolia (Allghoi Khorkhoi aka Mongolian death worm)

- 2006 Gambia (Gambo - Gambian sea monster , Ninki Nanka and Armitage's skink
- 2006 Llangorse Lake (Giant pike, giant eels)
- 2006 Windermere (Giant eels)
- 2007 Coniston Water (Giant eels)
- 2007 Guyana (Giant anaconda, didi, water tiger)
- 2008 Russia (Almasty)
- 2009 Sumatra (Orang pendek)
- 2009 Republic of Ireland (Lake Monster)
- 2010 Texas (Blue Dogs)
- 2010 India (Mande Burung)
- 2011 Sumatra (Orang-pendek)
- 2012 Sumatra (Orang Pendek)
- 2014 Tasmania (Thylacine)
- 2015 Tasmania (Thylacine)
- 2016 Tasmania (Thylacine)
- 2017 Tasmania (Thylacine)
- 2017 Russia (Almasty)
- 2018 Tajikistan (Gul)

For details of current membership fees, current expeditions and investigations, and voluntary posts within the CFZ that need your help, please do not hesitate to contact us.

The Centre for Fortean Zoology,
Myrtle Cottage,
Woolfardisworthy,
Bideford, North Devon
EX39 5QR

Telephone 01237 431413
Fax+44 (0)7006-074-925
eMail info@cfz.org.uk

Websites:

www.cfz.org.uk
www.weirdweekend.org

HOW TO START A PUBLISHING EMPIRE

Unlike most mainstream publishers, we have a non-commercial remit, and our mission statement claims that "we publish books because they deserve to be published, not because we think that we can make money out of them". Our motto is the Latin Tag *Pro bona causa facimus* (we do it for good reason), a slogan taken from a children's book *The Case of the Silver Egg* by the late Desmond Skirrow.

WIKIPEDIA: "The first book published was in 1988. *Take this Brother may it Serve you Well* was a guide to Beatles bootlegs by Jonathan Downes. It sold quite well, but was hampered by very poor production values, being photocopied, and held together by a plastic clip binder.

In 1988 A5 clip binders were hard to get hold of, so the publishers took A4 binders and cut them in half with a hacksaw. It now reaches surprisingly high prices second hand.

The production quality improved slightly over the years, and after 1999 all the books produced were ringbound with laminated colour covers. In 2004, however, they signed an agreement with Lightning Source, and all books are now produced perfect bound, with full colour covers."

Until 2010 all our books, the majority of which are/were on the subject of mystery animals and allied disciplines, were published by `CFZ Press`, the publishing arm of the Centre for Fortean Zoology (CFZ), and we urged our readers and followers to draw a discreet veil over the books that we published that were completely off topic to the CFZ.

However, in 2010 we decided that enough was enough and launched a second imprint, `Fortean Words` which aims to cover a wide range of non animal-related esoteric subjects. Other imprints will be launched as and when we feel like it, however the basic ethos of the company remains the same: Our job is to publish books and magazines that we feel are worth publishing, whether or not they are going to sell. Money is, after all - as my dear old Mama once told me - a rather vulgar subject, and she would be rolling in her grave if she thought that her eldest son was somehow in `trade`.

Luckily, so far our tastes have turned out not to be that rarified after all, and we have sold far more books than anyone ever thought that we would, so there is a moral in there somewhere…

Jon Downes,
Woolsery, North Devon
July 2018

CFZ PRESS

CFZ Press is our flagship imprint, featuring a wide range of intelligently written and lavishly illustrated books on cryptozoology and the quirkier aspects of Natural History.

CFZ Classics is a new venture for us. There are many seminal works that are either unavailable today, or not available with the production values which we would like to see. So, following the old adage that if you want to get something done do it yourself, this is exactly what we have done.

Desiderius Erasmus Roterodamus (b. October 18th 1466, d. July 2nd 1536) said: "When I have a little money, I buy books; and if I have any left, I buy food and clothes," and we are much the same. Only, we are in the lucky position of being able to share our books with the wider world. CFZ Classics is a conduit through which we cannot just re-issue titles which we feel still have much to offer the cryptozoological and Fortean research communities of the 21st Century, but we are adding footnotes, supplementary essays, and other material where we deem it appropriate.

http://www.cfzpublishing.co.uk/

Fortean Words is a new venture for us. The F in CFZ stands for "Fortean", after the pioneering researcher into anomalous phenomena, Charles Fort. Our Fortean Words imprint covers a whole spectrum of arcane subjects from UFOs and the paranormal to folklore and urban legends. Our authors include such Fortean luminaries as Nick Redfern, Andy Roberts, and Paul Screeton. . New authors tackling new subjects will always be encouraged, and we hope that our books will continue to be as ground-breaking and popular as ever.

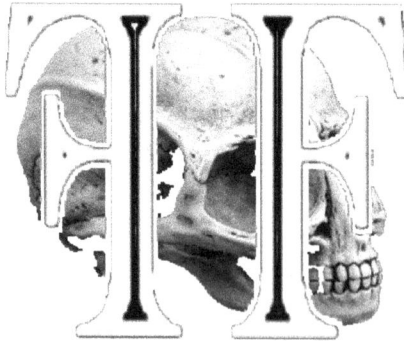

Just before Christmas 2011, we launched our third imprint, this time dedicated to - let's see if you guessed it from the title - fictional books with a Fortean or cryptozoological theme. We have published a few fictional books in the past, but now think that because of our rising reputation as publishers of quality Forteana, that a dedicated fiction imprint was the order of the day.

http://www.cfzpublishing.co.uk/

www.ingramcontent.com/pod-product-compliance
Lightning Source LLC
Chambersburg PA
CBHW072121270326
41931CB00010B/1630